氧化石墨烯与植物修复系统交互作用的研究

杜俊杰 著

中国农业出版社
北 京

前 言

环境中重金属和有机污染物的排放量长期保持比较高的水平,土壤污染也趋向严重化和复杂化。植物修复技术具有原位修复、环境友好、成本低和美化环境等优点,受到广泛的关注。然而植物修复技术周期长、效率低,适当的强化剂将更有助于植物修复技术的成功实施。石墨烯基纳米材料由于具有优异的物理化学性质被广泛用于各行各业。氧化石墨烯(Graphene oxide,GO)比表面积大、表面功能基团密度高,还具有良好的生物相容性和水溶液稳定性,在土壤污染治理领域可发挥重要作用。本书主要介绍了超积累植物在植物修复中的应用,以及氧化石墨烯在植物修复系统中与有关因子的交互作用,以期为石墨烯基纳米材料在污染土壤修复的应用提供研究案例和奠定一定的理论基础。

本书的设计、撰写和整理均由杜俊杰副教授完成,书中所述的实验研究是在课题组全体老师和同学的共同努力下完成的。特别感谢南开大学周启星教授、胡献刚教授、刘维涛副教授、欧阳少虎博士,山西财经大学侯萊博士,国家食品安全风险评估中心吴永宁研究员,陕西科技大学李国梁教授,山西师范大学刘维仲教授、徐建国教授、邰刚教授、吴建虎副教授、李桂峰副教授、陈伟副教授、李国琴博士和南阳师范学院高园园博士的支持和帮助。

在本书出版之际，感谢国家自然科学基金青年项目（31600411）、山西省基础研究计划面上项目（202103021224263）和山西师范大学博士启动基金的资助。

由于作者水平有限，书中难免有不足之处，敬请广大读者批评和指正。

杜俊杰

2022 年 6 月

目 录

前言

第1章 绪论 ·· 1
  1.1 植物修复 ·· 1
  1.2 影响植物修复的土壤因素 ··· 13
  1.3 Cd/POPs 污染土壤的植物修复手段 ································· 15
  1.4 纳米材料在植物修复中的应用 ·· 17
  1.5 纳米材料的生态毒性效应 ··· 19

第2章 土壤类型对植物修复 Cd 污染土壤的影响 ····················· 21
  2.1 引言 ·· 21
  2.2 盆栽实验方案与设计 ·· 22
  2.3 植物地上部生物量和 Cd 积累特征 ································· 27
  2.4 土壤特性指标和植物指标的相关性 ································ 29
  2.5 细菌菌群结果分析 ··· 30
  2.6 土壤特性指标和菌群指标的相关性 ································ 33
  2.7 小结 ·· 33

第3章 土壤类型对 PCBs 消散的影响 ······································· 35
  3.1 引言 ·· 35
  3.2 盆栽实验方案与设计 ·· 36
  3.3 4 种土壤类型的理化性状 ··· 39

3.4 土壤类型对 PCBs 消散的影响 ································ 40
3.5 不同类型土壤中酶活性对 PCBs 消散的响应 ············ 41
3.6 不同类型土壤中细菌菌群对 PCBs 消散的响应 ········ 44
3.7 小结 ·· 51

第 4 章 垃圾改良土壤对孔雀草及其根际微生物的毒性影响 ··· 53

4.1 引言 ·· 53
4.2 盆栽实验方案与设计 ··· 54
4.3 矿化垃圾和普通土壤的特性分析 ···························· 56
4.4 植物指标的测试结果 ··· 56
4.5 菌群分析结果 ·· 58
4.6 讨论 ·· 65
4.7 小结 ·· 66

第 5 章 氧化石墨烯对植物根部健康的影响研究 ················ 68

5.1 引言 ·· 68
5.2 水培实验方案与设计 ··· 69
5.3 植物根部生理和生化变化 ······································· 72
5.4 根部内生细菌菌群丰富度和多样性分析 ·················· 75
5.5 根部内生细菌菌群结构和表型分析 ························ 75
5.6 小结 ·· 81

第 6 章 植物根系分泌物对氧化石墨烯特性的影响 ············· 82

6.1 引言 ·· 82
6.2 水培实验方案与设计 ··· 83
6.3 根系分泌物鉴定结果 ··· 87
6.4 根系分泌物引发氧化石墨烯形貌的改变 ·················· 93
6.5 根系分泌物引发氧化石墨烯粒径分布和
    表面电荷的改变 ··· 95

目　录

6.6　根系分泌物引发氧化石墨烯表面化学官能团的改变 …… 96
6.7　根系分泌物引发氧化石墨烯表面化学性质的改变 …… 101
6.8　环境影响分析 …… 103
6.9　小结 …… 105

第7章　纳米材料对小麦种子萌发能力的影响研究 …… 106

7.1　引言 …… 106
7.2　萌发实验方案与设计 …… 107
7.3　纳米银粉和纳米银片对小麦种子萌发的影响 …… 108
7.4　单壁碳纳米管和多壁碳纳米管对小麦种子萌发的影响 …… 109
7.5　单层石墨烯和石墨烯纳米片对小麦种子萌发的影响 …… 110
7.6　小结 …… 111

第8章　氧化石墨烯在土壤中性质改变和引起的土壤细菌菌群变化 …… 113

8.1　引言 …… 113
8.2　土壤培育实验方案与设计 …… 114
8.3　土壤细菌菌群的丰富度和多样性 …… 117
8.4　土壤细菌菌群的分类组成 …… 119
8.5　氧化石墨烯表面形貌变化 …… 123
8.6　氧化石墨烯表面化学官能团的变化 …… 125
8.7　分析鉴定氧化石墨烯表面固定的小分子有机物 …… 129
8.8　氧化石墨烯粒径分布和表面电荷的变化 …… 132
8.9　氧化石墨烯化学活性和结构的变化 …… 134
8.10　小结 …… 136

第9章　石墨烯基纳米材料对土壤细菌菌群的影响研究 …… 138

9.1　引言 …… 138

9.2 土壤培育实验方案与设计 ………………………………… 138
9.3 石墨烯基纳米材料对土壤细菌菌群 α-多样性的影响 …… 140
9.4 石墨烯基纳米材料对土壤细菌菌群组成的影响 ………… 141
9.5 石墨烯基纳米材料对土壤细菌菌群功能组的影响 ……… 143
9.6 小结 ………………………………………………………… 150

第 10 章 氧化石墨烯对植物修复苯并[a]芘污染土壤的影响 ……………………………………………………… 151

10.1 引言 ………………………………………………………… 151
10.2 盆栽实验方案与设计 ……………………………………… 152
10.3 土壤 B[a]P 的消散以及植物对 B[a]P 的提取 ………… 154
10.4 土壤细菌菌群测序结果 …………………………………… 157
10.5 土壤细菌菌群 α-多样性分析结果 ……………………… 157
10.6 土壤细菌菌群结构分析结果 ……………………………… 159
10.7 讨论 ………………………………………………………… 162
10.8 小结 ………………………………………………………… 165

第 11 章 碳基负载纳米材料对植物修复复合污染初探 ……… 166

11.1 引言 ………………………………………………………… 166
11.2 盆栽实验方案与设计 ……………………………………… 167
11.3 纳米强化剂的 SEM 表征结果 …………………………… 171
11.4 纳米材料对孔雀草生长状况的影响 ……………………… 172
11.5 纳米强化剂对孔雀草生物量的影响 ……………………… 173
11.6 纳米强化剂对孔雀草提取 Cd 的影响 …………………… 174
11.7 纳米强化剂对土壤 PCBs 去除率的影响 ………………… 175
11.8 小结 ………………………………………………………… 177

参考文献 …………………………………………………………… 179

第 1 章 绪论

1.1 植物修复

植物修复是指利用植物把水或土壤基质中的有害外来物质固定、降解或提取出来的技术[1]。植物修复可以划分为五种类型：植物固定、植物提取、植物根际过滤、植物挥发和植物降解[2]。对于受污染的土壤，植物提取是一种最有效的植物修复方式。植物提取是利用超积累植物从被污染的土壤中提取重金属并将其富集和转移到相对容易处理的地上部分[3,4]。现在有文献记录的共有约 500 种超积累植物品种，占所有被子植物品种的 0.2%，其中大部分为 Ni 超积累植物[5,6]。但是，仍然有很多被报道的具有超积累重金属特性的植物品种还未被证实[7]。而且，许多被发现的超积累植物生长速度慢、生物量低，导致其植物修复效率很低[8]。因此，对于继续筛选超积累植物品种和探索新的筛选方法仍然很迫切。另外，提高植物修复效率还可以通过采用物理、化学和生物等方面的强化措施来实现。而且，土壤提供了超积累植物的生长基质，土壤-植物交互作用在超积累植物发挥修复作用过程中也起着关键的作用[9]。

1.1.1 超积累植物

"超积累植物"（Hyperaccumulators）由 Brooks 等在 1977 年最先提出，用来指在自然环境中生长的干重 Ni 含量超过 1 000mg/kg 的一类植物[10]。目前，超积累植物通常用来描述能主动从土壤中过量提取一种或几种微量元素的一类植物。如今，超过 500 种被子植物品种被鉴定为金属元素（Ni、Zn、Pb、Cd、Cr、Mn、Cu、Co、U、Sb 和 Ti 等）、类金属元素（As）和非金属元素（Se）的超积累植物[11]。与此同时，关于新的超积累植物品种的报道也不断出现[12,13]。然而，很多从污染区筛选得到的植物可能会从超积累植物品种名单中被删除，因为它们在受控条件的实验中，其特质可能会不符合超积累植物的评判标准[14]。

目前，评定超积累植物主要有 4 个指标：临界浓度、转移系数、富集系数、耐性特征[15]。另外，理想的超积累植物也应该具有其他一些特质。例如，抗病能力强、生物量大、生长周期短、超积累 1 种以上重金属等[16]。

1. 临界浓度标准 植物体重金属浓度的临界值通常用来衡量超积累植物的积累能力，超积累临界浓度因金属元素不同而不同，是指植物地上干物质中的重金属含量超过正常生理水平[5,17]。Brooks 等[10]首先用 1 000mg/kg 作为 Ni 超积累植物的临界值。后来，研究人员根据每种重金属元素的植物毒性特征选择一个临界含量作为针对该种重金属元素的超积累植物的判定标准[18]。Malaisse 等[19]以 1 000mg/kg 作为 Cu 超积累植物的临界含量。Reeves 等也以 1 000mg/kg 作为 Pb 超积累植物的临界含量[20]。Baker 等[5]建议以 10 000mg/kg 作为 Mn 和 Zn 的临界含量。也有一些学者建议，将植物地上部（茎和叶）重金属含量超过普通植物在同一生长条件下的 100 倍作为临界含量[21,22]。Wei 等[16,23,24]在前人研究的基础上并结合自己的研究结果，针对超积累植物临界含量标准给出了系统性建议：Zn、Mn 为 10 000mg/kg；Pb、Se、Cu、Ni、Co、As 等均为 1 000mg/kg；Cd 为 100mg/kg；Au 为 1mg/kg。

2. 转移系数标准 Chaney 等[21]提出，超积累植物必须能够高效地把重金属从根部转移至茎部。转移系数（Translocation factor，TF）是指植物地上部（主要是指茎和叶）重金属浓度与根部的比值大于 1，即 TF>1。转移系数标准用来衡量植物把其根部吸收的重金属元素转移到地上茎叶的能力。Salt 等[22]认为，超积累植物地上可收获部分的重金属元素浓度必须远高于土壤浓度，才能保证通过植物修复来实现污染土壤的重金属含量不断降低。安鑫龙等[25-27]从植物生理学角度考虑，提出超积累植物应满足细胞内矿质元素浓度大于细胞外矿质元素浓度。

3. 富集系数标准 富集系数（Bioaccumulation factor，BF）是指植物组织中积累的重金属浓度和所生长的土壤中重金属浓度之比，应该大于1，有时甚至可以达到50～100[28]。BF>1意味着植物地上部重金属浓度高于污染土壤的重金属浓度。因为植物对重金属的积累量随着土壤中重金属浓度的升高而增加，所以对超积累植物来说，这是一个关键标准[29,30]。当土壤中某种重金属浓度远大于临界浓度，非超积累植物体内的重金属浓度也有可能达到临界浓度。但是，一旦土壤中重金属浓度稍低于临界浓度，这些非超积累植物可能就跟普通植物一样，体内的重金属浓度将会低于临界浓度标准。所以，植物地上部较高的富集系数是超积累植物的必备特征，它可鉴别伪超积累植物。对于超积累植物，当土壤中重金属浓度与临界含量标准浓度相当时，植物地上部富集系数至少应该大于1[16]。

4. 耐性特征标准 毫无疑问，植物超积累重金属的基础是对高浓度重金属的耐性。在人为控制的试验条件下，与对照相比，受污染的超积累植物地上部生物量没有下降，甚至有所上升[23]。对于在自然污染状态下生长的超积累植物来说，是指从植物的生长状态来看，没有表现出明显的毒害症状，如组织坏死、萎蔫、发黄等[24]。相反，对于大多数普通植物来说，当污染物浓度高到一定程度，其生长就会受抑制，生物量明显下降。

1.1.2 超积累植物的解毒机制

对高浓度重金属元素的耐性，对于超积累作用来说是一个关键的植物特征要求，植物根、茎和叶细胞必须能耐受高浓度的重金属。这种耐受能力被认为是由于细胞外或细胞壁上的沉淀形成、细胞内区域化分布、螯合效应和酶系统等作用导致的。

1. 沉淀作用 在根系，超积累植物通过分泌根系分泌物，能够通过在细胞外形成磷酸镉、磷酸锌和硫酸铅等沉淀来抵抗重金属的毒害。此外，细胞呼吸产生的二氧化碳可能会在细胞外形成碳酸铅等沉淀。一般而言，不溶性的磷酸盐、碳酸盐和硫酸盐是金属胞外沉淀的主要形式，在植物解毒重金属过程中起到关键作用[31]。

2. 细胞区隔化 在超积累植物体内，有毒元素通常被区隔在对植物必需的细胞过程损害最小的区域。从组织水平看，重金属积累在植物的表皮层和毛状体；从细胞水平看，重金属在液泡和细胞壁积累[3,32]。Küpper 等[33]通过能量色散 X 射线微区分析技术研究 Zn 超积累植物 *Thlaspi caerulescens* 的叶片，发现其通过将 Zn 区隔在表皮细胞的液泡中来实现解毒作用。后来的研究还发现，Zn 超积累植物 *Arabidopsis halleri*（L.）在根表皮的细胞壁主要以磷酸盐沉淀的形式积累 Cd 和 Zn[34]。

3. 螯合作用 植物组织中的螯合剂在耐性机制、区隔分布机制、有机或无机污染物转移过程中都起很重要的作用[35]。超积累植物利用螯合剂与自由的重金属离子结合形成螯合物质，从而能够对体内过量的重金属元素起解毒作用，进而耐受重金属的毒性。据研究报道，超积累植物体内存在的主要金属螯合剂包括：金属硫蛋白、植物螯合肽、烟草胺、麦根酸类物质、有机酸（柠檬酸和苹果酸等）、谷胱甘肽、硫代葡萄糖苷、氨基酸（半胱氨酸和组氨酸）等，见表 1-1。

超积累植物体内的各种螯合剂对不同重金属元素的螯合作用各有特点。金属硫蛋白（MTs）和植物螯合肽（PCs）是超积累植物体内两种主要的金属螯合剂，都是含丰富半胱氨酸的多肽，MTs 可以高效结合 Cu，PCs 可以高效结合 Cd、Cu 和 As[7]。PCs 与 Cu 和 Cd 的螯合作用力很强，与 Zn 和 Al 亲和力很弱[36]。作为有机酸类螯合剂，与苹果酸相比，柠檬酸与 Zn 的螯合作用更强[37]。超积累植物体内生成的组氨酸通过含 N 配体与金属离子螯合形成复合物而实现解毒[38]。而超积累植物体内产生的谷胱甘肽存在多种解毒机制，既可以作为螯合剂与金属离子结合，也可以通过抗氧化作用减少金属元素对植物体造成的氧化损伤，还可以作为底物生物合成另一种螯合剂（PCs）[6]。

第1章 绪 论

表1-1 超积累植物体内常见的金属螯合剂

螯合剂	螯合元素	参考文献
金属硫蛋白（Metallothioneins，MTs）	Cu	Murphy et al[39]
植物螯合肽（Phytochelatins，PCs）	Cu、Cd As	Meharg et al[36] Pickering et al[40]
烟草胺（Nicotianamine，NA）	Cu、Zn、Fe、Mn	Stephan et al[41]
麦根酸类物质（Phytosiderophores）	Fe	Higuchi et al[42]
硫代葡萄糖苷（Glucosinolates）	Zn、Cd	Kusznierewicz et al[43]
谷胱甘肽（Glutathione，GSH）	Ni Co、Zn As	Freeman et al[44] Freeman et al[45] Wei et al[46]
半胱氨酸（Cysteine）	Cu Cd、As Ni	Robinson et al[47] Visoottiviseth et al[7] Krämer et al[6]
组氨酸（Histidine）	Zn Ni	Küpper et al[48] Kramer et al[38]
柠檬酸（Citrate）	Zn Ni Fe	Sarret et al[49] Lee et al[50] Von et al[51]
苹果酸（Malate）	Zn	Sarret et al[49]
硫酸盐（Sulfate）	As	Wei et al[46]

4. 酶作用 在一些植物组织内，特定酶参与的一些代谢活动可以通过改变污染物的物理状态或者化学成分而对重金属元素解毒。Freeman等[52]通过提高超积累植物 *Thlaspi goesingense* 体内的丝氨酸乙酰转移酶（Serine acetyltransferase，SAT）活性，从而提高了谷胱甘肽的含量，结果导致 Ni、Co 和 Zn 的耐受水平提

高。Benzarti 等[53]研究发现，当 *Thlaspi caerulescens* 暴露于高浓度的 Cd 条件下，植物体表观特征并未表现出异常，而其体内的谷胱甘肽还原酶（GR）和超氧化物歧化酶（SOD）的活性显著增强。酸性磷酸酶也是一种重要的重金属抗性酶，在细胞壁和液泡等部位合成来抵御重金属对植物的毒性[54]。还有研究发现，将小麦液泡的酸性磷酸酶基因 *TaVP1* 转移到烟草中并进行表达，烟草对 Cd 的耐受性增强，Cd 的积累量也增大[55]。

1.1.3　超积累植物的筛选

目前所报道的超积累植物大多都从地球化学异常区或重金属污染区筛选得到，包括金属采矿区、金属冶炼厂周边或者其他工业活动区的周边土地。研究发现，已知的 Zn 超积累植物品种都是从历史上的污染区周围和近代的采矿区筛选得到。例如，东南景天（*Sedum alfredii* H.），就是在浙江省衢州市周边的锌矿区发现的一种锌超积累植物。

关于超积累植物的筛选方法，仍在不断改良和完善。起初，研究人员曾认为只通过植物标本调查分析就能筛选超积累植物。Brooks 等[10]通过检测镍矿指示植物的标本中重金属含量发现了 5 种 Ni 超积累植物。之后，很多发表的超积累植物的筛选研究都是基于植物标本分析，却忽略了对土壤-植物交互作用的研究；筛选到的超积累标本中重金属含量均能达到临界含量标准，但其富集系数和转移系数是否满足标准不得而知[56,57]。

后来，研究人员在前人的研究基础上，结合野外调查和盆栽试验进行超积累植物的筛选。Ma 等[58]在美国佛罗里达中部一处被铬化砷酸铜（Chromated copperarsenate，CCA）污染的土地发现了蕨类植物蜈蚣草（*Pteris vittata* L.），并通过在实验室进行栽培实验验证了其是一种有效的 As 超积累植物。同时，陈同斌等也在我国境内发现了蜈蚣草[59]。除了上述植物品种，在我国境内的金属厂矿周边还筛选到了多种针对不同金属元素的超积累植物，见表1-2。

表 1-2 常见的在我国境内筛选的超积累植物品种

元素	超积累植物品种	发现地	参考文献
Zn	东南景天（Sedum alfredii H.）	浙江省衢州市周边的锌矿区	杨肖娥等[60]
As	蜈蚣草（Pteris vittata L.）	湖南省石门县石门雄黄矿矿区	陈同斌等[61]
As	大叶井口边草（Pteris cretica L.）	湖南省石门县石门雄黄矿矿区	韦朝阳等[62]
Cd	宝山堇菜（Viola baoshanensis）	湖南省郴州市桂阳县宝山矿区	刘威等[63]
Mn	商陆（Phytolacca acinosa Roxb）	湖南省湘潭市锰矿废弃的尾矿区	薛生国等[64]
Cu	鸭跖草（Commelina communis）	湖北省铜绿山古冶炼渣堆	束文圣等[65]
Cr	李氏禾（Leersia hexandra Swartz）	广西北部电镀厂周边	张学洪等[66]
Cd	龙葵（Solanum nigrum L.）	中国科学院沈阳生态实验站周边	魏树和等[67]

但是，在金属矿区周围生长的植物种群通常属于顶级群落。当人们从这些地区生长的各种植物中筛选超积累植物时，一些具有超积累特性的土著植物品种、先驱物种和中间物种在演替过程中有可能会被忽略。而且，一些被认定的超积累植物也可能是在污染地区因长期适应而形成的没有超积累基因的普通植物[16]，这些伪超积累植物也可以被称为耐重金属植物，它们在污染区和未污染区都可以生长，但只在污染区具有超积累能力[68]。基于以上考虑，Wei 等[16]提出，在正常的未污染区也可能存在重金属超积累植物，并通过田间盆栽模拟试验调查和研究了中国科学院沈阳生态实验站周边 20 科 54 种杂草品种，发现了一种新的 Cd 超积累植物——龙葵。在加入 25mg/kg Cd 的土壤中生长的龙葵，其茎和叶中积累的 Cd

浓度分别达到 103.8mg/kg 和 124.6mg/kg[12]。在未污染区筛选到超积累植物在筛选方法上是一种尝试和突破。而且，田间盆栽模拟试验筛选超积累植物的优点在于：①盆栽植物与自然生长的植物生长条件很接近；②可以完整观察候选植物长期的生长过程，有利于鉴定植物品种；③相对于在采矿区污染土壤长期暴露的植物来说，从未污染区选择盆栽植物可以大大缩短污染暴露时间。未污染区的超积累植物体内可能含有与超积累功能相关的基因，那将是非常有用的植物修复资源。而采矿区的植物具有的超积累特征可能是一系列环境诱导的生理适应能力[69]。

1.1.4 超积累植物修复效果的强化措施

利用超积累植物对重金属污染的土壤进行植物修复的优点有：原位修复、环境友好、成本低、美化环境。但是，植物修复技术的局限性体现在周期长和效率低，适当的强化措施将更有助于植物修复技术的成功实施。

1.1.4.1 物理方法

物理修复技术，例如玻璃化、电动力技术、蒸汽提取技术、热解析和超声等，都已在重金属污染土壤修复实践中得以应用[70]。因为这些技术要求专门的设备和消耗能量，对土壤的物理结构和生物有效性都有不利的影响，所以这些技术成本高而且对环境不友好。但是，有些区域有土地利用的压力或者巨大的发展潜力，有时污染物对人类健康或环境有急迫危险性，遇到这些情况，物理修复技术可以作为植物修复的补充措施[71]。

1.1.4.2 化学强化

据研究报道，常用于活化重金属和提高植物对重金属积累量的化学强化剂主要是一些金属螯合剂，包括：EDTA（Ethylenediaminetetraacetic acid）、HEDTA（Hydroxyethyl ethylenediamine triacetic acid）、CDTA（Cyclohexane-diamine-tetraacetic acid）、DTPA（Diethylenetriaminepentaacetic acid）、EGTA（Ethylene glycol tetraacetic acid）、EDDHA［Ethylenediaminedi-（o-hydroxyphenylacetic) acid]、EDDS (S, S-ethylenediaminedisuccinic acid)、NTA (Nitrilotriacetic acid) 和 CA (Citric acid) 等[72,73]。向土壤中加入这些螯合剂可以把金属引入土壤溶液中,将土粒吸附的金属解析出来,例如溶解铁锰氧化物、溶解沉淀等。而且,金属-螯合剂复合物还可以抑制金属发生再次沉淀和被吸附,进而提高植物提取的生物可利用性[31]。Marques 等[74]以 *Solanum nigrum* 从土壤提取 Zn 为模型,比较了 EDTA 和 EDDS 两种螯合剂的强化效果,EDTA 和 EDDS 分别使 *Solanum nigrum* 叶片中 Zn 浓度提高至原来的 231%和140%。Quartacci 等[75]研究发现,与 NTA 相比,EDDS 从土壤解析 Cu、Pb 和 Zn 的能力更强,使其在 *Brassica carinata* 地上部中含量提高了 2~4 倍。可见,化学螯合剂确实可以强化超积累植物提取土壤中重金属的能力,但有一点应该注意,施用生物不可降解的螯合剂也会带来负面影响,例如对土壤微生物和植物根膜的毒性、污染地下水等[72]。

腐植酸是自然界广泛存在的天然物质,可以在植物修复中作为化学螯合剂的一种替代物。有研究发现,在污染土壤中添加 2g/kg 腐植酸,植物 *Nicotiana tabacum* SR-1 地上部 Cd 含量从 30.9mg/kg 提高至 39.9mg/kg[76]。腐植酸分子上的酸性官能团,如羧基和酚羟基,在重金属的运输、溶解性和生物利用性等方面起着重要作用[77]。而且,腐植酸可以降低土壤中重金属的物理流动,从而限制其渗滤到地下水,有助于环境保护[78]。另外,其他的化学强化剂,例如絮凝剂(海藻酸钠、淀粉、CMC、PAAS、PVP 等)、淋洗剂(HCl、H_2SO_4、HNO_3 等)和吸附剂(活性炭、蒙脱石、碳纳米材料等),这些化学添加剂的施加都将可能有助于超积累植物修复重金属污染的土壤。但是,在应用这些化学强化剂之前,应该首先考虑可能对环境带来二次污染的风险。

1.1.4.3 生物措施

增强植物修复效率的生物措施主要包括:扩大根系面积、施加微生物菌剂、植物基因工程技术等[79]。菌根真菌对植物有很多积极的影响,可以促进植物营养提取,为植物生产细胞分裂素和赤霉素等激素,这些积极的作用都有助于植物修复[77]。Baum 等[80]通过接种外生菌根菌 *Paxillus involutus* 提高了毛枝柳（*Salix dasyclados*）对 Zn 的提取量。Whiting 等[81]发现,在灭菌土壤上生长的 *Thlaspi caerulescens* 比在自然土壤生长时,地上部 Zn 积累量显著降低,说明土壤微生物对超积累植物提取重金属有重要作用。Belimov 等[82]从生长在 Cd 污染生境的 *Indian mustard* 根际分离出 11 株耐 Cd 细菌,其中大部分菌株体内都含有 1-aminocyclopropane-1-carboxylate（ACC）酶,这种酶能刺激植物根的伸长。可见,重金属离子对植物激素合成具有抑制作用。有研究发现,链霉菌（*Streptomyces* spp.）可以通过生成铁载体蛋白螯合金属离子,进而促进植物合成激素,强化植物修复[83]。Sheng 等[84]研究还发现,从土壤分离的耐 Cd 菌株培养在含 $CdCO_3$ 的培养基中,可以检测到吲哚乙酸。所以,微生物可以通过直接或间接作用产生植物激素,提高植物修复重金属污染的效率,因为植物激素有助于提高植物地上部生物量和根部伸长,从而提取更多的重金属。

近几年,在植物对重金属超积累和超耐性的分子机制方面取得了突出的进展。拥有耐重金属、积累重金属和解毒重金属能力的植物和微生物为这种特殊的基因提供了一个重要的基因库,可以将这些基因转移到生长周期短和生物量高的植物品种而提高植物修复效率[85]。微生物的多样性和适应能力使其能够在高等植物不能生长的严峻的有毒环境中生存,微生物作为一个重金属解毒功能基因库,将其在超积累植物体内高效表达,可能组装出更高效的植物修复品种[86]。

目前,关于超积累植物的转基因技术研究主要通过基因修饰使金属运载蛋白、金属螯合剂和金属同化反应关键酶在植物体大量合成,强化植物对重金属的提取能力。ATP 硫化酶（APS）在植物吸收同化 Se 的代谢过程中起关键作用,ATP 硫化酶基因在 Indian mustard（*Brassica juncea*）体内的高度表达增强了其对 Se 的耐性和积累能力[87]。研究人员还通过将金属离子转运蛋白基因 *NtCBP*4 和 *CAX*2 分别转移到烟草（*Nicotiana tabacum*）体内进行高度表达，提高了烟草对金属离子的耐受能力和积累能力[88,89]。通过转基因，金属螯合剂，例如金属硫蛋白（MTs）[90]、植物螯合肽（PCs）[91]、柠檬酸盐（CA）[92]、铁蛋白（Ferritin）[93]等，也都成功在植物体内大量合成，强化了植物对重金属的提取能力。但是，转基因植物也有可能因生物入侵而带来生态风险。叶绿体基因工程技术（Chloroplast genetic engineering）提供了一种获得高度表达目的基因的新方法，这种技术不会把目的基因通过花粉传播到邻近的野草和作物，生态风险较小[94-96]。

除了微生物菌剂和转基因植物，蚯蚓和跳虫等土壤无脊椎动物的引入对超积累植物修复污染土壤的效果也具有强化作用。Ma等[97]研究了蚯蚓（*Pheretima guillelmi*）对银合欢（*Leucaena leucocephala*）修复广东省某 Pb/Zn 尾矿土壤效果的影响，发现蚯蚓的引入使植物生物量显著提高，重金属的提取量也提高了16%～53%。研究人员在研究蚯蚓（*Dendrobuena rubida*）对英国威尔士重金属污染土壤溶解性的影响时，发现蚯蚓粪便中水溶态 Pb 的含量比土壤高 50%[98]。蚯蚓和跳虫等土壤动物的生命活动可改善土壤理化性状、增强土壤养分循环、促进植物生长、提高植物生物量、提高土壤微生物数量和活性；还能通过取食、消化和排泄等生命活动以及与微生物的相互作用提高土壤中重金属的生物有效性[99]。可见，蚯蚓等土壤动物也能通过直接或间接的作用强化植物修复效果。

1.1.4.4 农艺强化

正常情况下，植物修复要求植物在污染地种植很长周期，因此，必要的农艺措施有助于使植物生物量和重金属积累量最大化。农艺措施主要包括施肥和耕作技术等，在植物修复过程中很重要[1,100]。如果在了解农艺实践和管理知识的基础上采取一些适当的强化措施，可以提高植物修复的效率[101]。

土壤 pH 是决定植物积累重金属量的一个重要因素。对大部分元素而言，降低 pH 可以打破溶解-沉淀平衡体系，促进土壤溶液中重金属的释放[101]。所以，通过向土壤中施加酸性肥料可以增加植物对重金属的提取量，例如硫酸铵[102]。与此相反，土壤溶液中 As 含量会因 pH 降低而减少释放，加入氧化钙这样的碱性物质可以有效提高 As 的植物提取量。当然，像化学螯合剂一样，加入化学肥料同样也带来环境风险[103]。所以必须采取安全的操作过程和保护措施避免二次污染。此外，加入化学营养肥料，如尿素、钾肥、磷肥和复合肥等，都能促进植物生长，提高生物量，进而增加植物的重金属提取量。

有机肥也有能促进植物生长和提取重金属，例如畜禽粪便和堆肥污泥。加入有机肥可以增加土壤有机质含量，N、P 和 K 元素含量，微生物数量，提高脲酶和磷酸酶活性[104]。Wei 等[72]发现，施用鸡粪虽然使超积累植物 *Bidens tripartite* L. 的 Cd 含量有所降低，但其植物生物量提高 4 倍多，最终增强了对 Cd 的提取效率。Sung 等[105]研究发现，施用堆肥为植物营造松散透气的土壤结构，还有助于溶解土壤金属沉淀和为植物提供营养，促进植物生长和提取重金属。

作物栽培技术也可以促进植物生长，增加植物生物量，提高植物提取效率和缩短修复周期。作物栽培技术可以考虑的方面有很多，包括种植方法、种植密度、田间除草、作物收获时间和方法、种子加工和控制授粉等，这些耕作管理措施的良好控制都有助于提高植物提取效率[106]。

种子包衣和幼苗移栽，有助于提高污染地播种的成活率。种子包衣技术是指在种子外边包裹一层肥料和杀虫剂，提高发芽率和抑制种子病虫害。利用特殊合成的微生物菌剂作为包衣材料，可以实现超积累植物和微生物联合修复，大大提高重金属的植物提取效率。直接移栽幼苗可以缩短植物修复周期[101]。

由于植物在苗期和花期对水和肥敏感，所以适当施肥将有助于提高超积累植物地上部生物量[107]。Wei 等[108]通过一年两次在花期收获植物的方法提高了 *Rorippa globosa* 去除土壤 Cd 的效率。Ji 等[109]也通过研究证明，选择一个最佳的收获时间和收割位置将有助于提高 *Solanum nigrum* L. 对土壤 Cd 的提取量。另外，种植密度对植物生物量和重金属累积量有影响。因为种植过密，植株竞争营养和水分，降低植物修复效率。McGrath 等[100]认为，生育时期对 *Thlaspi caerulescens* 提取 Zn 和 Cd 的效率也影响显著。环境因子，如光照、温度、水、空气和热量等，都明显影响植物生长，通过调控环境因子可缩短超积累植物的生长周期，提高植物修复效率[110]。何振立[111]认为，提高 Eh 可以增加土壤溶液中重金属离子浓度。农艺措施，如增加日晒、间歇灌溉和作物轮作等，都可以调节土壤 Eh[101]。

1.2　影响植物修复的土壤因素

土壤是一个巨大的综合体，不仅是植物生长的基质，也可以过滤降水和废水，既产生气体也吸附气体，是有机体的大本营。土壤在超积累植物形成过程中起了关键的作用，其影响土壤中金属的生物可利用性和植物生长状况等。

1.2.1　土壤粒径

土壤是一种由固相、液相和气相组成的分散系。从粒径上分，土壤颗粒又包括沙粒、粉粒和黏粒。沙粒和粉粒是由岩块破碎而来，它们在化学性质上比黏粒相对稳定，是相对不活泼的。土壤粒径分布决定土壤的质地，直接影响土壤的物理化学和生物学性质，与植物生长所需的环境条件和养分转化关系密切。土壤粒径越小，其吸湿能力和吸水能力越强，对超积累植物根系的水分供应影响很大。有机质和细黏粒是金属污染土壤中主要的金属负载体，也是超积累植物的营养物质来源[112]。只有了解土壤颗粒的组成和质地特性及其与土壤肥力的关系，才能采取适当的措施对不良质地土壤加以改良，为植物修复提供一个良好的生活环境。在农业生产实践中，对土壤的要求既要保水又能通气，既能吸水又能供水，既容易耕作又不能散成单粒或结成大块。因此，必须考虑各种颗粒的合理搭配，理想的土壤应该是沙粒、粉粒及黏粒的一种混合体系，各种颗粒的特性兼而有之。

1.2.2 土壤质地

为了便于描述土壤质地,需要使用一些特殊的名字,例如,沙壤土、粉壤土、沙质黏土等。沙土含水量低,热容量小,春季升温快,所以超积累植物在沙土上种植发苗也快。相反,在黏土上种植超积累植物发苗较慢。壤质土的特性介于黏土和沙土之间。一般而言,比较好的土壤质地应该是包含 10%~20% 黏土,沙粒和粉粒含量接近,而且含有一定量的有机质。

1.2.3 土壤比表面积

土壤颗粒的比表面积对植物修复也是很重要的,因为许多物理和化学反应在土壤颗粒表面发生。在植物修复过程中,大量的反应,例如污染物解读、吸附和解析,生物降解等,都在颗粒表面进行。土壤颗粒越小,比表面积越大。

1.2.4 土壤孔隙度

土壤孔隙度过大不利于根系固定,容易失水;孔隙度过小根系不易下扎、伸展。孔隙度过大或过小都对超积累植物根系与污染物的接触不利。一般而言,最适宜的土壤孔隙度应该是 8%~10%。

1.2.5 土壤温度

超积累植物多被发现于温带和热带地区[5]。气候带直接影响空气温度和土壤温度,土壤温度和空气温度对植物生长都很重要。土壤表面温度随不同季节与不同时间段波动和变化。土壤被稠密的植物或厚层覆盖物所掩盖,温度变化并不剧烈。土壤温度直接影响超积累植物生长和根系微生物活性变化。土壤冻和融也影响土壤的结构。缓慢的冻融变化对土壤结构有利,有利于植物生长。土壤温度决定了生化过程的方向和速率,了解土壤的温度变化对调节土壤热状况，提高土壤肥力，满足作物对土壤温度的要求都有重要的意义。

1.2.6 土壤颜色

土壤表层颜色可以反映许多土壤的特质，例如营养状况、土壤温度、土壤结构和土壤类型等，这些特性都直接影响超积累植物的生长。土壤表面的颜色主要由有机质含量决定，颜色越黑有机质含量越高。有机质赋予土壤很多优良的特质，例如，良好的聚合性、较高的持水能力等。黑色的土壤也能在白天吸收更多的光辐射，在夜里释放热量。土壤底层的颜色反映土壤的湿度和气体含量。一般而言，微红色和褐色的底层颜色表示较好的通气状况和较低的含水量。浅灰色、橄榄色指示较多的水分和铁的化学还原反应；浅灰色和褐色的杂色表示地下水位波动。

1.2.7 土壤类型

土壤类型包含了很多土壤信息，例如，主要的金属元素、土壤形成因素和土壤形成过程等。土壤类型的高级分类包括：土纲、亚纲、土类、亚类，低级分类包括土属、土种、变种。分类水平越低，土壤类型越具体，土壤信息越丰富。超积累植物的生长状况也能反映土壤类型。大部分超积累植物生长在四种类型土壤[113,114]：①蛇纹岩土壤：富含 Ni、Cr 和 Co；②菱锌矿土壤：Pb 和 Zn 含量丰富，Cd、As 和 Cu 含量也很高；③富 Se 岩风化土壤；④富含 Co 和 Cu 的土壤，由含金属硫化物的白云岩和黏土岩发育而成。

1.3　Cd/POPs 污染土壤的植物修复手段

土壤不仅是生态系统的重要组成，还是污染物的主要汇集地。随着我国工农业生产的迅速发展以及城镇化水平的不断提高，土壤环境污染问题日趋严重。多方面的数据资料[115-119]显示，重金属和持久性有机污染物（POPs）已经成为土壤环境污染的罪魁祸首，不仅影响了土壤的结构和性质，而且破坏了土壤的生态系统，更为严重的是这些污染物往往以作物或农业生态系统为中介，通过食物链把污染物传递到人类这一最高营养级，对人体健康的威胁日益严重。

目前，在土壤重金属污染中，以镉（Cd）、铅（Pb）和汞（Hg）最为突出。与国外相比，我国土壤 Cd 污染状况十分严重。最新报道[120]指出，我国农田土壤 Cd 污染面积已超过 $20 \times 10^4\ hm^2$，每年生产 Cd 含量超标的农产品达 $14.6 \times 10^8\ kg$。Cd 作为一种蓄积性毒物，可以通过食物链进入生物体并产生生物放大作用，其毒性是潜在的。研究表明：人体内 Cd 的半减期是 20～40 年[121]。此外，Cd 不是植物生长的必需元素。土壤中 Cd 含量过高将对植物的生理生态有多方面的不良影响，而且对细胞的毒害具有明显的生物累积效应，破坏农作物的正常生长和遗传功能。为了解决土壤的重金属污染问题，常规的治理方法是采用工程措施或化学措施，但成本均较为昂贵，而且还会破坏土壤结构及微生物区系，也容易引起"二次污染"。目前，在土壤 Cd 污染的治理、修复方面，利用植物进行生态修复是研究的热门，其原理是利用某些对 Cd 具有超、高富集能力的植物将土壤中的 Cd 大量地转移到植株体内从而达到修复土壤的目的[122]。这种途径修复潜力大，而且可以维持土壤肥力，保持土壤结构和区系生物群落免遭破坏，还能营造良好的生态环境。近年来，我国在 Cd 超积累植物的筛选、性能强化和应用研究方面进展迅速，取得了不少研究成果[123,124]。虽然植物修复是近年来公认的理想的污染土壤原位修复技术，但由于目前所发现的大多数超积累植物存在生物量小、生长缓慢、适应能力差、对有机-无机复合污染修复能力差等缺点，因此，大大限制了植物修复的发展空间。孔雀草（*Tagetes patula* L.）作为一种较为理想的 Cd 超积累植物，是近年来从众多花卉植物中筛选出来的。该植物为菊科万寿菊属，一年生草本花卉，具有很好的观赏价值，与其他 Cd 超积累植物相比，它既耐移栽，又生长迅速，栽培管理比较容易；撒落在地上的种子在合适的温、湿度条件下可自行生长，具有抗逆性和适应性强等特点，在世界各地广泛种植。这些特点无疑有利于其在污染土壤植物修复中广泛应用。

多卤代芳烃化合物（PHAHs）和多环芳烃（PAHs）都是土壤环境中典型的 POPs。PHAHs 包括多氯联苯（PCBs）、多氯苯并二噁英（PCDDs）、多氯二苯并呋喃（PCDFs）和 2，3，7，8-四氯苯并二噁英（TCDD）等。它们被广泛应用于电容器、塑料、油漆、油墨、无碳纸和除草剂等的生产中，并可通过各种途径进入环境中。PAHs 包括萘、菲、蒽、荧蒽、芘和苯并[a]芘（B[a]P）等。PAHs 是重要的环境和食品污染物，具有潜在的致畸性、致癌性和基因毒性，该类化合物具有极低的水溶性，在环境中很难降解，正呈不断增加的趋势[125,126]。环境中的 POPs 由于受气候、生物、水文和地质等因素的影响，在不同的环境介质间发生一系列的迁移转化，最终贮存在土壤、河流和沿岸水体的底泥中。POPs 进入土壤容易被有机质吸附，难以横向和纵向迁移，部分 POPs 可通过挥发进行地—气交换，大多数 POPs 长期存在于土壤表层难以自然降解，因此土壤是 POPs 的主要"汇"[127]。针对土壤 POPs 污染这一问题，国内外开展了相关处置与治理方法的研究[128]，目前主要有：物理法、化学法、生物降解法和植物修复法。虽然 POPs 处理方法很多，但每种方法都存在着局限性。物理和化学修复技术成本昂贵，不易大面积推广应用；微生物修复技术需要较高的 POPs 浓度，才能维持群落的稳定，并且特定微生物只能降解特定 POPs；植物修复方法容易受到修复植物根际所达范围和生长速度的局限。

1.4 纳米材料在植物修复中的应用

纳米材料是一种新兴材料，从 20 世纪 80 年代才被开始研究。与普通材料相比，表面效应、体积效应和久保效应赋予了纳米材料特有的性能,使其在电子、计算机和医药等方面被日益广泛应用。在环境保护方面,纳米材料利用自身巨大的比表面积、强的催化和吸附能力等对污染水体和土壤进行处理,并表现出极佳的处理效果。目前,纳米铁是处理污染水体和土壤的一种典型纳米材料[129]。与纳米碳管、纳米TiO_2、纳米ZnO和纳米螯合剂相比,纳米铁易得、成本低,并且可以还原和催化环境中多种污染物,在污染环境修复方面将表现出越来越大的优势[130]。

自20世纪80年代,Sweeny等[131]首次报道了金属铁还原氯代脂肪烃的研究后,零价铁由于来源丰富,价格便宜,并且对某些有机氯化合物能够迅速还原脱氯,因此金属零价铁还原脱氯技术被广泛重视。然而,随着研究的深入,研究者发现该技术的实施还存在着很大的问题:金属铁处理低氯化物时反应活性较低,易造成氯化产物的生成和累积;随着反应时间的增加,金属铁表面由于形成了一层钝化膜使得反应活性降低。针对上述不足,国内外学者又研究出改进的零价铁修复技术,包括纳米铁、在铁表面镀一层贵金属的双金属体系铁的化合物等。

石墨烯是由碳原子紧密堆积而成的单层二维蜂窝状碳原子晶体,其晶体薄膜厚度只有0.355nm,因具有独特的物理化学性质,可与污染物之间形成非常强的络合能力,因而可以吸附污染物[132,133]。多氯联苯优先聚集在小颗粒上,并且大颗粒(直径>63μm)主要吸附低氯代的多氯联苯,而高氯代的组分吸附于小颗粒上(直径<63μm)。氧化石墨烯(GO)是石墨烯基纳米材料的一种,其表面含有大量的含氧活性基团如羧基、羰基、羟基和环氧基等,其应用涉及环保、能源和生物医药等多个领域[134]。GO不仅比表面积大,表面功能基团密度高,还具有良好的生物相容性和水溶液稳定性。在治理环境有机污染物方面,目前已有报道GO可以有效吸附水环境中的萘、菲和芘等多种PAHs[135,136]。GO表面带负电荷,呈酸性,通过π-π堆叠、氢键和阴阳离子反应等作用力与有机污染物结合[137]。但目前仍缺乏利用GO对污染土壤的修复

研究，更没有把 GO 与植物修复技术联合运用到污染土壤的修复中。石墨烯基材料在植物修复系统中与各环境因子的交互作用是值得探讨的问题。

1.5 纳米材料的生态毒性效应

近年来，纳米材料在多个领域得到广泛研究和应用，随之，纳米材料的环境排放量也将大大增加[138]。如今正处于纳米产业发展的初期，在纳米材料大批量释放到环境之前，应该仔细评估其对生态环境的影响[139]。其中，粮食作物作为初级生产者，纳米材料的谷物毒性尤其值得关注。纳米材料是指在三维空间中至少有一维处于纳米尺度范围（1~100nm）或由它们作为基本单元构成的材料[140]。常见的纳米材料类型包括：碳基纳米材料（石墨烯、碳纳米管和富勒烯等）、金属基纳米材料（纳米金属单质或者氧化物）和复合材料等。据报道，纳米材料对植物种子萌发能力具有促进作用还是抑制作用，仍然没有形成统一的认识。Siddiqui 等[141]研究发现，纳米 SiO_2 促进番茄种子萌发和早期幼苗的生长。Lin 等[142]研究认为，纳米 ZnO 抑制玉米种子发芽和幼苗根伸长。还有研究发现，多壁碳纳米管对萝卜种子萌发没有影响，而纳米 Ag 却使萝卜种子发芽率降低 20%，根伸长减少 70%[143]。

2012 年科学技术部将纳米研究纳入国家科技发展战略，2013 年成立"中国石墨烯产业技术创新战略联盟"，随后常州一家年产 100t 氧化石墨烯/石墨烯生产线投产；2013 年氧化石墨烯/石墨烯类纳米材料入选欧盟"未来新兴旗舰技术项目"，并获得 10 亿欧元资助。与此同时，美国、日本、新加坡等国家也在密集支持氧化石墨烯/石墨烯的研发和应用。相比石墨烯，氧化石墨烯具有合成简单、易于后期修饰加工、水相分散性好等优点，作为卫生医学、化学化工、电子产品、家用电器、汽车飞机、环境保护技术等的关键功能材料正在迅速进入人们的生产生活中[144,145]。随着氧化石墨烯的大量生产及广泛应用，其不可避免地释放到生态环境当中，考虑其显著的纳米效应、较强的环境稳定性及突出的反应活性，氧化石墨烯引发的生态风险不容忽视[146]。近几年，GO应用领域和应用产品的迅猛增长必将导致高水平的环境排放[147]。GO给生态环境所带来的影响需要在其大量释放之前得以调查研究[148]。

第 2 章 土壤类型对植物修复 Cd 污染土壤的影响

2.1 引言

重金属可通过施用化学肥料、污泥、堆肥和农药等途径进入土壤，并在土壤中蓄积，现已成为一个严峻的生态问题[149,150]。在重金属元素中，镉（Cd）是生态毒性较强的一种，对土壤生物活性、食品安全和人类健康都具有负面影响[151]。可用于重金属污染土壤修复的技术有多种，包括物理和化学技术、电化学技术、植物修复技术和微生物修复技术等[152]。但是，土壤修复的最终目标不仅是只将金属元素从土壤中去除，还应该能提高土壤质量[153,154]。作为植物修复技术中的一种方法，植物提取技术利用超积累植物从土壤中提取金属元素并转移到容易处理的地上部，具有环境友好和成本低廉等优点[155]。而且，有研究表明，植物修复手段还能改善土壤质量[153,156]。迄今为止，全世界超过 500 种超积累植物已被发现[157]。其中，有很多 Cd 超积累植物，例如玉米、芥菜、向日葵、蚕豆、三叶鬼针草、龙葵等[158-160]。一般情况下，不建议利用可食农作物进行植物修复，防止它们将重金属带进食物链，给人类健康造成威胁[161]。其实，观赏性植物用于植物修复是一个很好的选择。孔雀草（*Tagetes patula*）就是一种具有重金属超积累特性的园艺植物，还具有抗逆性强的优点[162]。

虽然超积累植物在提取重金属方面具有很多优点，但其缺点也是存在的，例如，很多报道的超积累植物都具有地域性[163]。植物成功提取土壤重金属主要依赖于其本身对金属元素的积累能力、植物品种和土壤条件[164]。土壤条件影响植物生物量、重金属的生物可利用性和微生物菌群。如今，关于土壤类型对孔雀草提取 Cd 效率的影响研究较少，关于土壤各种特性对植物提取效率的贡献研究也较少，但这些问题对植物提取技术的广泛应用至关重要。

土壤微生物菌群影响土壤质量、土壤修复和土壤生态功能[160,165]。而且，土壤菌群结构和组成也受土壤性质和植被的影响[156]。土壤微生物菌群通过代谢调控土壤各元素的生物地球化学循环，被认为是土壤健康的一个关键指标[166]。Cd 对几乎所有的细菌都有毒性[167]。Cd 的存在会对微生物菌群造成生存压力，菌群会逐渐做出反应和改变。所以，弄清植物修复系统中土壤微生物菌群非常重要。而且，不同土壤类型及其特性对土壤菌群的影响也很少报道。与传统的平板培养和 PCR-DGGE 等技术相比，高通量测序技术是一种调查土壤细菌菌群的最好选择[168]。

在本研究中，通过对不同类型土壤人工染毒进行盆栽试验，调查不同土壤类型及土壤特性对植物提取效率和土壤细菌菌群的影响[169]。

2.2 盆栽实验方案与设计

2.2.1 土壤预处理

4 种土壤的采集坐标分别为 47°17′N、132°35′E（黑土）、31°22′N、120°44′E（水稻土）、43°58′N、81°31′E（绿洲土）和 39°11′N、117°01′E（潮土），采土深度为 0～20cm。土壤样品先经风干，再过筛（2mm 筛孔）去除大颗粒杂质，最后按照常规方法测定理化性状[170]。

（1）土壤 pH。pH 可影响土壤肥力，不仅影响土壤的属性、组成以及土壤中物质的存在形式和有效性，而且影响土壤微生物的活动和植物的正常生长。称取 10g 土壤于 50mL 塑料离心管中，然后加入 25mL 去 CO_2 水，恒温振荡 3min 后离心，最后用 pH 计测定 pH。

第 2 章 土壤类型对植物修复 Cd 污染土壤的影响

(2) 阳离子交换量（CEC，cmol/kg）。用 1mol/L 中性乙酸铵溶液（pH 7.0）反复处理土样，致使土壤中 NH_4^+ 饱和。多余的乙酸铵用乙醇洗去之后，再用去离子水把土样洗入凯氏瓶，然后加氧化镁固体蒸馏。用硼酸溶液吸收蒸馏出来的氨气，最后用盐酸标准溶液进行滴定。根据 NH_4^+ 的量计算阳离子交换量[171]。

(3) 有机质（OM，g/kg）。在加热条件下，用过量的硫酸-重铬酸钾溶液氧化土壤中的有机碳，用硫酸亚铁标准溶液对多余的重铬酸钾进行滴定，利用消耗的重铬酸钾量按氧化校正系数计算得到有机碳量，最后乘以常数 1.724，即为土壤有机质含量。

(4) 全氮（TN，%）。土样中的亚硝态氮经高锰酸钾氧化后转化为硝态氮，然后用铁粉还原硝态氮并转化为铵态氮。在有加速剂参与的条件下，经浓硫酸消煮，各种含氮有机化合物转化为铵态氮。用硼酸吸收蒸馏出来的氨，并用酸标准溶液进行滴定，计算得出土壤全氮含量。

(5) 全磷（TP，%）。土样经氢氧化钠熔融后，含磷矿物和有机磷化合物都转化为可溶性的正磷酸盐，在酸性条件下正磷酸盐与钼锑抗显色剂发生反应而生成磷钼蓝，然后在波长 700nm 处测定吸光度，吸光度值与样品中总磷的含量符合朗伯-比尔定律。

(6) 全钾（TK，%）。土样中的有机物先经高锰酸钾和硝酸加热氧化而除去，然后用氢氟酸分解硅酸盐等矿物，生成四氟化硅而逸去。继续加热赶酸，使矿质元素转变为盐类或金属氧化物。所得残渣用盐酸溶液溶解，致使残渣中的钾转变为钾离子。经定量稀释后用原子吸收分光光度法测定钾离子浓度，最后换算成土样中全钾的含量。

(7) 粒径分布（MD，%）。本研究使用 Mastersizer 2000 激光粒度仪测定土粒平均直径（体积加权平均直径）。激光粒度仪的工作原理是基于光学中的米氏散射理论和夫朗霍夫衍射理论，建立由偏振光源（激光光源）、粒子通路和检测系统三部分组成的激光粒度分析仪光路系统。

土壤中含有大量的有机质、碳酸钙等，会影响粒度测量的真实结果。因此用激光粒度分析仪测定土壤粒径分布，需要在上机操作之前对样品进行预处理，去除有机质、钙质等成分，以及胶结作用对粒度的影响。

取土样 0.5～0.8g 放入 100mL 烧杯中，加入 10mL 浓度为 10% 的过氧化氢煮沸，使其充分反应以去除有机质；然后加入 10mL 浓度为 10% 的盐酸煮沸，使其充分反应去除碳酸钙。然后反复加蒸馏水静置，其间不断吸出上清液，直至呈中性，抽去蒸馏水，加入 10mL 浓度为 0.05mol/L 的六偏磷酸钠分散剂后上机测量。参数设置：测量范围 0.02～2 000μm；转速：2 000r/min；超声：240W。

不同土壤类型的理化性状在表 2-1 中列出。按照染毒浓度要求（10mg/kg），通过将 $CdCl_2 \cdot 2.5H_2O$ 标准品配制成标准溶液加入土壤样品中，并充分混匀。取 1.5kg 染毒土置于花盆中平衡 30d 备用。

表 2-1 土壤理化性状

指标	黑土	水稻土	绿洲土	潮土
pH	5.84	6.26	7.78	7.74
阳离子交换量（CEC，cmol/kg）	17.40	18.00	5.88	4.25
有机质（OM，g/kg）	33.00	47.00	22.60	8.18
粒径（MD，μm）	8.16	9.77	18.22	20.51
总氮（TN,%）	0.162	0.230	0.168	0.068
总磷（TP,%）	0.94	1.80	2.61	1.05
总钾（TK,%）	0.35	0.48	0.40	0.16
染毒后真实 Cd 浓度（mg/kg）	9.98	9.96	9.93	9.92

注：表中数值均为平均值（$n=3$）。

2.2.2 盆栽试验

孔雀草种子购于北京花儿朵朵花卉种子公司。种子先经 5% H_2O_2 浸泡 20min 后用纯净水反复清洗，再置于铺有双层湿纱布的托盘上发芽。种子发芽后转移到装有洁净土的育苗盘，放在人工气候箱进行培养（25℃）。孔雀草幼苗长出4~6片真叶后，转移到装有染毒土的花盆进行盆栽试验[162]。盆栽条件：25℃，50%田间持水量，90d。每种土壤类型的盆栽试验都做3个重复。盆栽试验结束后，收集土壤和植物样品，用于后续的重金属和微生物分析。

2.2.3 Cd分析

将收获的植物分为地上部（茎叶）和地下部（根）两部分，用自来水反复冲洗，并将根部放进20mmol/L Na-EDTA溶液中浸泡15min，以除去可能吸附在根表面的金属离子。然后用去离子水冲洗，沥去水分，在−55℃条件下冷冻干燥至恒重，粉碎并过筛（2mm筛孔）备用。定量称取粉碎后的植物样品0.200 0g于消解罐中，加入2mL H_2O_2 和6mL HNO_3，装入微波消解仪中密封消解。消解程序：120℃、8min；150℃、6min；180℃、6min。消解液用3% HNO_3 定容后，经0.45μm水系滤膜过滤，上机（ICP-MS）测定Cd含量。

土壤样品经冷冻干燥至恒重、研磨、过筛（2mm筛孔）后，定量称取土样0.200 0g于消解罐中，先后加入5mL HNO_3、2mL H_2O_2 和2mL HF，装入微波消解仪中密封消解。消解程序：150℃、10min；180℃、10min；200℃、40min。消解完全后，放在智能控温电加热器蒸至近干（防止HF腐蚀玻璃仪器），然后用3% HNO_3 定容，经0.45μm水系滤膜过滤，上机（ICP-MS, Elandrce, USA）测定Cd浓度。

富集系数（BF）由植物地上部Cd浓度与土壤Cd浓度（Ce）的比值计算而得。植物地上部Cd重量由地上部Cd浓度与地上部干重相乘计算而得。

2.2.4 土壤细菌菌群分析

分析方法主要参考之前发表的论文，根据实际需求有所改动[172,173]。主要步骤包括：DNA提取、PCR反应、高通量测序、序列加工处理和生物信息学分析。

1. DNA 提取和 PCR 反应　称取 0.5g 新鲜土样进行 DNA 提取，按照 Fast DNA SPIN extraction kits（MP Biomedicals，Santa Ana，CA，USA）试剂盒的说明进行操作。通过 PCR 反应扩增细菌 16S rRNA V3～V4 区的 DNA 片段。引物（primer）为 338F（5′-ACTCCTACGGGAGGCAGCA-3′）和 806R（5′-GGACTACHVGGGTWTCTAAT-3′）。

反应条件如下：

预变性：98℃，2min；25 次循环：98℃变性 15s，55℃复性 30s，72℃延伸 30s。

延伸：72℃，5min。

反应体系（20μL）：

5×Q5 reaction buffer	5μL
5×Q5 High-Fidelity GC buffer	5μL
Q5 High-Fidelity DNA Polymerase（5U/μL）	0.25μL
2.5mM dNTPs	2μL
Forward primer（10μM）	1μL
Reverse primer（10μM）	1μL
DNA Template	2μL
ddH$_2$O	8.75μL

2. 高通量测序和生物信息学分析　用 2%琼脂糖回收扩增产物，按照 Agencourt AMPure Beads（Beckman Coulter，Indianapolis，IN）试剂盒的说明书纯化扩增产物，最后按照 PicoGreen dsDNA Assay Kit（Invitrogen，Carlsbad，CA，USA）试剂盒的说明书进行定量。合并后的扩增产物在 Illumina MiSeq 平台进行 Pyrosequencing 高通量测序（上海美吉生物医药科技有限公司）。原始序列已经上传至 NCBI Sequence Read Archive（SRA）数据库（Accession Number：PRJNA531289）。

使用软件 QIIME（version 1.8.0）进行质控处理序列，低质量序列筛除标准如下：

(1) 去除长度<150bp 的 reads。
(2) 去除 Phred 平均分数<20 的 reads。
(3) 去除含有模糊碱基和含 2 个以上错配引物碱基的 reads。
(4) 去除 overlap 超过 10bp 的 reads。

对于有效序列，使用 UPARSE（version 7.1）将相似度超过 97%以上的归类到一个 OTU（operational taxonomic units），利用 Uchime（http：//www.mothur.org/wiki/Chimera.uchime）去除 chimeric sequences。Sequence data 分析主要利用 QIIME 和 R packages（version 3.2.0）进行。使用 BLAST 并比对 Greengenes 数据库，对基因序列进行系统发育分析，置信范围为 70%。

通过 RDP 数据库（SSU115）将基因序列系统分类到门（phylum）、纲（class）、目（order）、科（family）和属（genus）5 个水平。通过 MOTHUR 获得每个样品的 Shannon 指数。利用 R Project 软件进行层序聚类分析。

2.2.5 数据分析

所有实验都进行 3 次重复，所得的数据均进行方差分析，误差棒代表了标准差，结果以 SPSS 19.0 统计软件进行分析，数据的显著水平均指 $P<0.05$（ANOVA，Tukey）。土壤理化性状指标和其他分析指标之间的相关性分析通过 Pearson 相关性分析方法在 SPSS 19.0 软件计算而得。

2.3 植物地上部生物量和 Cd 积累特征

植物地上部生物量和 Cd 浓度直接决定了植物提取 Cd 的效率。如图 2-1 和图 2-2A 所示，孔雀草干重由高到低依次为：绿洲土（7.32g/盆）>黑土（6.99g/盆）>水稻土（5.86g/盆）>潮土（1.01g/盆）。潮土的植物干重显著低于其他土壤（$P<0.05$）。如图 2-2B 所示，孔雀草 Cd 富集系数（BF）由高到低依次为：黑土（9.23）>水稻土（7.44）>绿洲土（5.45）>潮土（5.40）。BF>1

表明孔雀草地上部 Cd 浓度高于土壤 Cd 浓度，说明孔雀草能从此 4 种土壤中高效提取 Cd，并转移到地上部。黑土的富集系数显著高于绿洲土和潮土（$P<0.05$）。植物的修复效率依赖于植物干重和 Cd 浓度的乘积。如图 2-2C 所示，孔雀草地上部 Cd 提取量由高到低依次为：黑土＞水稻土＞绿洲土＞潮土。黑土的 Cd 提取量显著高于绿洲土和潮土（$P<0.05$）。

图 2-1　孔雀草生长状况

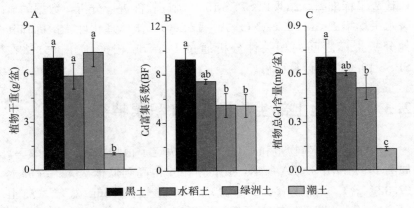

图 2-2　孔雀草生物量和 Cd 积累能力

注：误差棒代表平均值±标准差，不同小写字母表示在 0.05 水平（Tukey 检验）差异显著性，$n=3$。

由此可见，土壤类型影响植物生物量和 Cd 积累浓度，并最终影响植物修复效率。Zehra 等[164]也发现，与始成土（美国土壤分类系统）相比，向日葵在氧化土（美国土壤分类系统）中生长状况更好，Cd 富集能力更强。

2.4 土壤特性指标和植物指标的相关性

通过 Pearson 相关性分析评价各种土壤特征对植物提取 Cd 效率的贡献，结果见表 2-2。BF 与土壤 pH 显著负相关（$P<0.05$），而和其他土壤性质没有显著性关系。但是，BF 与 CEC、OM、TN 和 TK 呈不显著的正相关关系；BF 与 MD 和 TP 呈不显著的负相关关系。孔雀草干重与 CEC、OM、TN、TP 和 TK 呈不显著的正相关关系；孔雀草干重与 pH 和 MD 呈不显著的负相关关系。植物地上部 Cd 重量直接反映植物修复效率。综合来看，孔雀草地上部 Cd 重量与 CEC、OM、TN、TP 和 TK 呈正相关，与 pH 和 MD 呈负相关。众多土壤特征指标中，pH 对孔雀草 Cd 富集能力的影响最强。

表 2-2 土壤特性与植物指标的相关性

指标	pH	CEC	OM	MD	TN	TP	TK
富集系数（BF）	−0.97*	0.90	0.66	−0.95	0.46	−0.49	0.38
植物干重	−0.45	0.52	0.65	−0.59	0.77	0.50	0.83
植物地上部 Cd 重量	−0.77	0.81	0.83	−0.86	0.83	0.18	0.82

注：CEC，阳离子交换量；OM，有机质含量；MD，平均粒径；* 显著性水平 $P<0.05$。

土壤 pH 越高，Cd 的生物可利用率越低，植物提取 Cd 的能力越弱[174]。Yang 等发现土壤 pH 下降 0.5 个单位，Cd 的可利用浓度增加 2 倍[175,176]。土壤 Cd 的生物可利用性顺序由高到低依次为：可交换态＞碳酸盐态＞铁锰氧化态＞有机态＞残渣态[177]。可交换态很容易被植物转移和吸收。碳酸盐态、铁锰氧化态和有机态在较低 pH 条件下才可能被植物利用。残渣态相对稳定，不能被植物吸收[178]。Karna 等发现，重金属的生物可利用率随土壤粒径的减小而增加[179]。有机质可作为螯合剂促进 Cd 在土壤中的移动性，促进植物提取。CEC 越高，植物对营养物质的提取能力越强[180]。尽管磷酸盐能固定金属元素而降低其生物可利用性，但土壤较高的 P 元素也能促进植物生长和发育[181]。在本研究中，P 含量与 BF 呈负相关，与孔雀草生物量呈正相关，但最终与孔雀草地上部 Cd 重量呈正相关。

2.5 细菌菌群结果分析

高通量测序结果见表 2-3，各实验组的 OTU 反映菌群丰富度，由高到低依次为：水稻土＞黑土＞绿洲土＞潮土。Shannon 指数反映菌群多样性，由高到低依次为：黑土＞水稻土＞绿洲土＞潮土。

表 2-3 菌群丰富度和多样性指数

土壤类型	Reads	3% distance	
		OTU	Shannon
黑土	17 827	1 080±185	5.90±0.12
水稻土	23 726	1 148±164	5.84±0.25
绿洲土	17 029	861±91	5.78±0.02
潮土	15 316	730±127	5.09±0.06

根据系统分类学结果，如图 2-3 所示，4 种土壤的优势菌群在门水平上是相似的。对照组的洁净土壤中，9 个优势菌群（Acidobacteriota、Chloroflexi、Proteobacteria、Actinobacteriota、Firmicutes、Gemmatimonadetes、Bacteroidetes、Myxococcota 和 Desulfobacterota）的丰度分别占黑土、水稻土、绿洲土和潮土的 95.97%、92.83%、94.35%和 94.58%。处理组土壤中，8 个优势菌群（Proteobacteria、Acidobacteria、Chloroflexi、Firmicutes、Actinobacteria、Planctomycetes、Gemmatimonadetes 和 Bacteroidetes）的丰度分别占黑土、水稻土、绿洲土和潮土的 91.78%、

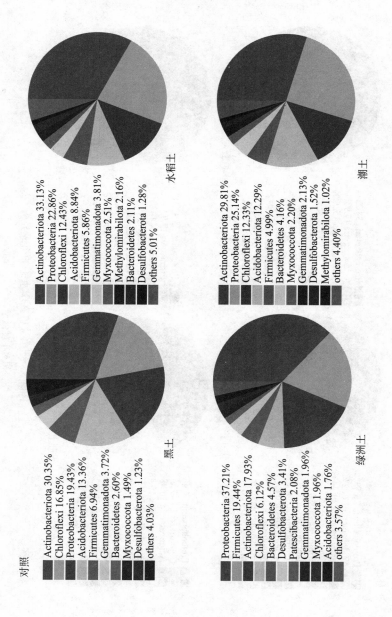

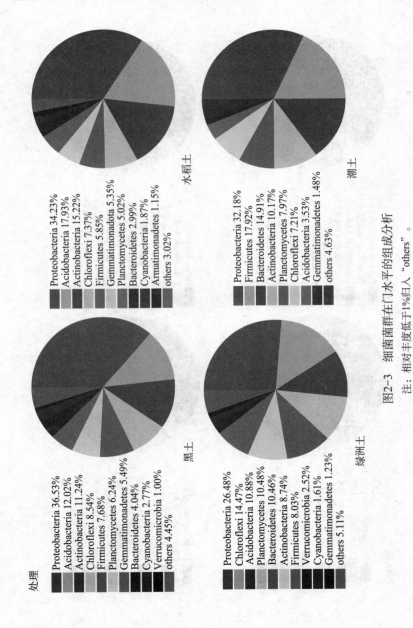

图2-3 细菌菌群在门水平的组成分析

注：相对丰度低于1%归入"others"。

93.96%、90.75%和95.37%。经过植物修复处理，Myxococcota 和 Desulfobacterota 不再是优势菌群，而 Planctomycetes 成为新的优势菌群。无论是对照组还是处理组，4 种土壤的菌群丰度都有所差异。从以上结果可见，不管是否经历植物修复过程，4 种土壤中的优势菌群大体相似，只是各种优势菌群的丰度不同。而且，经过植物修复，菌群结构也发生小程度的改变。

尽管 4 种土壤的特性不同，但菌群响应是相似的。Cd 污染土壤中存在的优势菌群对 Cd 具有耐性。重金属对微生物的毒性表现在抑制细胞代谢和能量代谢。在 Cd 胁迫压力存在下，耐 Cd 菌群被选择出来[182]。有关利用高粱进行植物修 Cd 污染土壤的研究也发现了相似的优势菌群[156]。

2.6 土壤特性指标和菌群指标的相关性

曾有研究认为，细菌菌群的变化与土壤特征的改变有关联[183,184]。本研究中（表 2-4），OTU 与 CEC 显著正相关（$P<0.05$），与 MD 显著负相关（$P<0.05$）。Shannon 指数与任何土壤理化指标都没有显著性关系。Hoshino 等研究也认可这一点[185]。以上结果表明，土壤 MD 和 CEC 对 4 种土壤的细菌菌群丰度有重要影响。

表 2-4 土壤特征和菌群指数的相关性

指标	pH	CEC	OM	MD	TN	TP	TK
OTU	−0.93	0.98*	0.94	−0.98*	0.82	−0.07	0.77
Shannon	−0.65	0.73	0.82	−0.77	0.87	0.36	0.89

注："CEC"，代表阳离子交换量；"OM"，代表有机质含量；"MD"，代表土粒直径；"TN"，代表总氮含量；"TP"，代表总磷含量；"TK"，代表总钾含量；* 显著性水平 $P<0.05$。

2.7 小结

孔雀草对 Cd 的提取效率由高到低依次为：黑土＞水稻土＞绿

洲土＞潮土。结合 4 种土壤的理化性状分析，土壤 pH 对孔雀草 Cd 富集量影响最大。土壤细菌菌群多样性由高到低依次为：黑土＞水稻土＞绿洲土＞潮土。4 种土壤中的优势菌群相似，但丰度有差异。经植物修复后，土壤菌群结构发生了小的改变。土壤粒径与细菌菌群丰度呈负相关，而 CEC 与细菌菌群丰度呈正相关。由此可推断，适当调整土壤理化性状将有助于提升植物修复重金属污染土壤的效率，这也有助于植物修复技术在更大范围的应用。

第 3 章 土壤类型对 PCBs 消散的影响

3.1 引言

多氯联苯（Polychlorinated biphenyls，PCBs）是一种持久性有机污染物（Persistent organic pollutants，POPs），在环境中分布广泛[186,187]。PCBs 具有高毒性、难降解性和致癌性，长期威胁着人类健康和土壤生态功能[188,189]。在真实环境中发现的 PCBs 通常以复杂混合物的形式存在，这些混合物来源于商业配方，如 Aroclor 系列[190-192]。很多种类的工业产品都含有 PCBs，包括增塑剂、绝缘液和润滑剂等。这些工业品因使用不当、废物处理或者意外泄漏等原因很容易把 PCBs 排入环境[193,194]。虽然 PCBs 产品已被禁止生产 50 年，但其在环境样品中仍然可以检测出来[195,196]。PCBs 的化学稳定性使其能长期在土壤中存在，严重影响土壤的基本生态功能[197]。有研究认为，PCBs 在自然环境中可能会经历吸附、挥发、光解和化学降解等过程，微生物降解是环境对 PCBs 自净的主要过程[188,198]。PCBs 可以作为自然界微生物的唯一碳源或能源而被其降解。PCBs 降解菌通常属于有限的门（如 Chloroflexi[199]、Firmicutes[200] 和 Proteobacteria[198]）和属（如 *Pseudomonas*[201]、*Burkholderia*[202]、*Sphingomonas*[203] 和 *Rhodococcus*[204,205]）。土壤酶在催化有机物降解过程中起到重要的作用。土壤酶的活性可以反映土壤条件的变化和土壤的自净能力，可以作为微生物活性的一个指标[206,207]。

影响 PCBs 生物可利用性及其降解微生物的因素都可以影响 PCBs 降解过程。Ti 等在研究中通过偏最小二乘回归分析法，发现 PCBs 生物可利用性与土壤湿度和粒径呈正相关关系[208]。Lehtinen

等认为,污染物较高的生物可利用性与较低含量的阳离子交换量、有机质、Fe 和 Al 有关[209]。Starr 等认为,土壤有机碳对 PCBs 的生物可利用性影响最大[210]。土壤有机质含量显著影响 PCBs 的吸附/解吸过程,从而决定其生物可利用性。从另一个方面来说,土著微生物菌群的组成、结构、丰度、多样性和活性等也与 PCBs 降解有关[183,188]。土壤类型不同,其物理、化学和生物学特性都有差异。很多研究都发现,土壤理化性状对土壤微生物菌群的组成和功能具有重要影响[185,211,212]。因此,探求土壤类型对 PCBs 消散及其相关的酶活性和微生物菌群是非常重要的课题[213]。

在本研究中,通过对 4 种不同类型土壤人工染毒(Aroclor1242)进行土培实验 90d。通过分光光度法分析了 2 种水解酶(蛋白酶和磷酸酶)和 3 种氧化还原酶(过氧化氢酶、脱氢酶和漆酶)的活性变化。通过高通量测序技术分析了土壤细菌菌群的变化。通过以上实验和分析,目的是调查土壤类型对 PCBs 消散的影响以及消散过程中相关酶活性和微生物菌群的影响。

3.2 盆栽实验方案与设计

3.2.1 土壤预处理和实验设计

采集黑土、水稻土、绿洲土和潮土用于本研究,采土过程和土壤常规理化性状的测定同第 2 章。土粒表面形貌通过扫描电子显微镜(LEO-1530VP,Germany)观察测定。

PCBs 选择 Aroclor1242(AccuStandard Inc.,99%),Aroclor1242 标准品经正己烷溶解后加入土壤。有机溶剂在排气室挥发 72h 后与土壤充分混匀,最终浓度为 5mg/kg(干土)。在混匀过程中采取连续梯度逐级混合的方法,保证污染物与土样混合均匀。污染土在 25℃温室中平衡 2 周后装盆,每盆 1.5kg 土。灭菌水加入土样使最大田间持水量维持在 50%,培养室维持日间温度在 25℃,时间为 90d。相同重量的洁净土作为对照组,对照组和处理组在培养过程中的操作步骤保持一致。所有的实验组都做 4 个重复。培养结束后，部分土壤样品贮藏于−80℃用于DNA提取，剩下的样品用于PCBs含量测定和酶活性测定。

3.2.2 PCBs分析过程

冷冻干燥后的土壤样品经粉碎后过筛（0.15mm），定量称取20g样品并加入回收率指示物PCB209（AccuStandard Inc.，99%），在平衡24h后，用滤纸包裹后置于平底烧瓶中，加入200mL提取溶剂（丙酮：正己烷=1：1），在平底烧瓶中加入5片铜片（$1cm^2$）脱硫，放在水浴锅（59℃）中，控制回流速度5～6次/h，连续索氏抽提24h。

将提取液在旋转蒸发仪上（温度39℃）浓缩至1mL，而后溶剂置换成正己烷，接着继续浓缩至1mL，完成溶剂替换。采用硅镁吸附剂（Florisil）层析柱净化。采用湿法装柱法，层析柱长20cm，内径10mm。层析管先经正己烷（10mL）淋洗，从下端排出淋洗液，至液面刚好浸没石英砂层，废弃此淋洗液。关上开关，再加入20mL正己烷，用药匙将10g Florisil从顶端漏斗加至19cm的记号线位置。为了防止Florisil黏附在管壁上，在装柱过程中需要不断加入正己烷，并保持正己烷以一定的流速持续流出；为了防止层析柱出现断裂，必须保持正己烷液面始终高于层析柱界面；在装柱过程中，需要用吸耳球不断轻轻敲打层析柱，致使Florisil紧实。为了除去样品中的水分，最后以上述方法装入1cm无水硫酸钠。调节液面，使正己烷正好处于无水硫酸钠层面之上，关闭开关备用。用长颈滴管将浓缩液转移至净化柱中，浸泡10min，以便样品与净化柱充分接触交换。以20mL正己烷分3次洗涤盛装浓缩液的平底烧瓶，均用长颈滴管转移到净化柱中，然后以大约90滴/min的速度流出，并继续添加正己烷至洗脱溶剂总体积为150mL，用250mL平底烧瓶收集全部洗脱液。洗脱液再经浓缩至1～2mL，加入内标五氯硝基苯（AccuStandard Inc.，99% purity），用轻柔的氮气吹至近干，用色谱纯正己烷定容至0.2mL，上机（GC-MS）测定分析。

GC-MS（Agilent 6890N-Agilent 5973）参数：分离用的毛细管柱为 DB-5MS 柱（J&W，30m×0.25mm×0.1μm），载气为氦气，稳定的气流速度为 2mL/min，分离柱升温程序：起始温度 90℃（保持 0.8min），然后以 15℃/min 的速度升至 180℃，以 1℃/min 升到 220℃；最后以 7℃/min 升到 290℃（保持 2min），无分流进样（1μL）。不同样品的回收率分别为：78%～104%（黑土）、95%～102%（水稻土）、84%～108%（绿洲土）和 88%～95%（潮土）。

3.2.3 土壤酶活性分析过程

本研究根据前人的研究方法通过分光光度法测定土壤酶活性。利用 H_2O_2 作为底物测定过氧化氢酶活性，以 $\mu mol/(g \cdot d)$ 来表示[214]。利用 2,3,5-氯化三苯基四氮唑（2,3,5-Triphenyl Tetrazolium Chloride，TTC）为底物测定脱氢酶活性，以 $\mu g/(g \cdot d)$ 来表示[215]。利用 2,2'-联氮-双-3-乙基苯并噻唑啉-6-磺酸（2,2'-azinobis-3-ethylbenzothiazoline-6-sulfonic acid，ABTS）为底物测定漆酶活性，以 $nmol/(g \cdot min)$ 来表示[216]。利用酪蛋白为底物测定蛋白酶活性，以 $mg/(g \cdot d)$ 来表示[217]。利用对硝基苯磷酸二钠（p-nitrophenol phosphate disodium，PNPP）为底物测定磷酸酶活性，以 $\mu mol/(g \cdot d)$ 来表示[206]。在测定蛋白酶和磷酸酶的过程中，需要根据土壤 pH 选择适当的缓冲液。所有的测定都做 3 次重复。

3.2.4 细菌菌群分析过程

分析方法主要参考发表的论文，根据实验需要有所改动[173,218]。主要步骤包括：DNA 提取、PCR 反应、高通量测序、序列加工处理和生物信息学分析。详细步骤见第 2 章。原始序列已经上传至 NCBI Sequence Read Archive（SRA）数据库（Accession Number：PRJNA691684）。通过 RDP 数据库（SSU115）将基因序列系统分类到门（phylum）、纲（class）、目（order）、科

(family) 和属 (genus) 5 个水平。通过 MOTHUR 获得每个样品的 ACE 指数和 Shannon 指数。利用 R Project 软件进行层序聚类分析、文氏图绘制和 PCoA 分析。

3.2.5 数据分析

所有实验都进行 4 个平行处理，所有的测定都重复 3 次。所得的数据均进行方差分析，误差棒代表了标准差。利用 SPSS 19.0 进行结果分析，数据的显著水平均指 $P<0.05$。通过 Tukey 检验比较 4 种土壤类型之间的差异性（ANOVA，Tukey）；通过 t 检验比较同一种土壤的两种不同处理之间的差异性（ANOVA，t-test）。

3.3 4 种土壤类型的理化性状

4 种类型土壤的理化性状在表 3-1 中列出，不同位置采集的样品存在明显差异。根据 pH、CEC、SOM 和 MD 数据，4 种土壤可以分为 2 组。黑土和水稻土都是酸性土，阳离子交换量和有机质含量高于绿洲土和潮土；绿洲土和潮土都是碱性土，都具有较大的土粒直径。4 种土壤中，水稻土的有机质、TN 和 TK 含量最高；绿洲土的 TP 含量最高；潮土的土粒直径最大，SEM 图也证实了这一点（图 3-1）。而且，潮土的粒径分布也比其他土壤类型更均匀。潮土的有机质含量低可能是造成其粒径均匀的原因。水稻土和绿洲土中没有检出 PCBs。黑土和潮土的 PCBs 初始浓度分别为 $8.53\mu g/kg$ 和 $10.87\mu g/kg$。根据国家土壤环境质量评价标准，4 种土壤 PCBs 初始浓度均低于 $200\mu g/kg$，可以认为是未被 PCBs 污染的土壤[219]。

表 3-1 土壤理化性状

指标	黑土	水稻土	绿洲土	潮土
pH	5.84±0.13a	6.26±0.05a	7.78±0.16b	7.74±0.25b
CEC (cmol/kg)	17.40±0.36a	18.00±0.25a	5.88±0.15b	4.25±0.13c

（续）

指标	黑土	水稻土	绿洲土	潮土
OM（g/kg）	33.00±2.21a	47.00±4.53b	22.60±1.19c	8.18±0.78d
MD（μm）	8.16±0.32a	9.77±0.15a	18.22±0.38b	20.51±1.46c
TN（%）	0.162±0.005a	0.230±0.006b	0.168±0.016a	0.068±0.002c
TP（%）	0.94±0.04a	1.80±0.16b	2.61±0.33c	1.05±0.03a
TK（%）	0.35±0.05a	0.48±0.02b	0.40±0.02a	0.16±0.02c
PCBs（μg/kg）	8.53±0.79a	ND	ND	10.87±0.94a

注：表中数值均以"平均值±标准差"列出，同一行不同字母代表具有显著差异（$P<0.05$）。"PCBs"，代表所采集土壤的 PCBs 含量；"CEC"，代表阳离子交换量；"OM"，代表有机质含量；"MD"，代表土粒直径；"TN"，代表总氮含量；"TP"，代表总磷含量；"TK"，代表总钾含量；"ND"，代表没有检出。

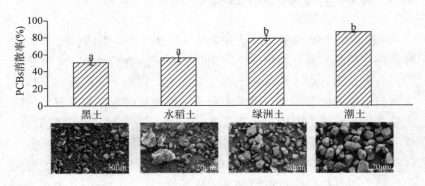

图 3-1 4 种土壤的 PCBs 消散率和颗粒形貌

注：误差棒代表平均值±标准差（$n=4$），不同字母代表具有显著性差异（$P<0.05$）颗粒形貌。

3.4 土壤类型对 PCBs 消散的影响

Aroclor 1242 的消散率结果见图 3-1，4 种土壤的消散率由高到低依次为：潮土（86.41%）＞绿洲土（79.31%）＞水稻土（56.09%）＞黑土（50.65%）。与土壤理化性状一致，4 种土壤的

PCBs 消散率也可以划分为两组。绿洲土和潮土的 PCBs 消散率接近，差异不显著（$P>0.05$），但与其他两种土壤的消散率差异显著（$P<0.05$）。水稻土和黑土的 PCBs 消散率差异不显著（$P>0.05$）。有研究认为，土壤理化性状可以影响有机污染物的生物可利用性[220]。土壤粒径越小，比表面积越大，导致对 PCBs 具有较高的吸附率和较低的生物可利用率[208]。有机质是影响有机污染物在土壤中蓄积的主要因素，有机质含量越高，污染物的保持力越强[171]。阳离子交换量和 pH 也影响有机污染物的生物可利用性和消散。Guimarães 等[221]研究发现，敌草隆（除草剂）在阳离子交换量高的土壤中降解率低。Zhang 等[222]研究发现，pH 在 DDT（杀虫剂）的消散过程中起着关键作用。有研究证明，五氯苯酚和芘在酸性土壤中的吸附率高于碱性土壤[223,224]。以上结果说明，具有不同特性的土壤类型对 PCBs 消散起关键的作用。

3.5 不同类型土壤中酶活性对 PCBs 消散的响应

土壤酶对环境压力反应迅速。水解酶和氧化还原酶是两类在生物修复污染物过程中非常重要的酶[225]。氧化还原酶中过氧化氢酶、脱氢酶和漆酶在分解外来有机化合物过程中具有重要的作用[226]。水解酶中蛋白酶和磷酸酶参与元素生物地球化学循环，在有机物转化和矿化过程中起着重要的作用[227]。

土壤类型和 PCBs 污染对土壤酶活性的影响，见图 3-2 和表 3-2。不同类型土壤的 5 种酶活性对 PCBs 污染的响应各不相同。与未污染的对照相比，黑土的磷酸酶、蛋白酶、脱氢酶和漆酶的活性显著升高（$P<0.05$）；水稻土的磷酸酶和漆酶活性显著降低，蛋白酶活性显著升高（$P<0.05$）；绿洲土的蛋白酶和过氧化氢酶活性显著降低（$P<0.05$）；潮土的蛋白酶、脱氢酶和漆酶活性显著升高（$P<0.05$）。对于未污染的对照组而言，每种酶的活性在不同类型土壤之间大体都呈现显著性差异（$P<0.05$）。染毒处理组的 5 种酶活性在不同土壤之间的差异变化与对照组一致。从以上结

果可以看出，不管土壤是否受到 PCBs 污染，土壤酶活性在 4 种类型土壤之间的差异保持相对稳定。这说明土壤类型对土壤酶活性的影响强于 PCBs 污染。

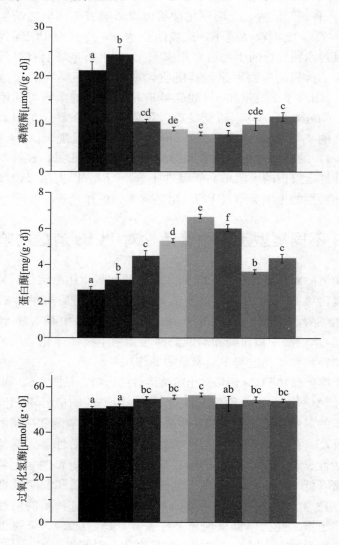

第3章 土壤类型对 PCBs 消散的影响

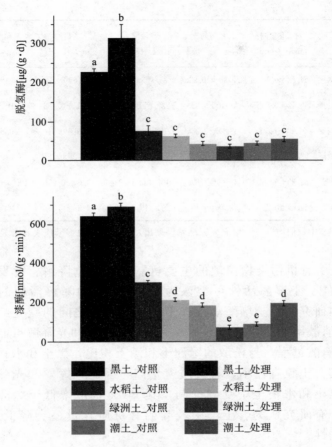

图 3-2 4种土壤在不同处理条件下的酶活性柱状图

注：误差棒代表平均值±标准差，不同小写字母表示在 0.05 水平差异显著，$n=4$。

表 3-2 4种土壤在不同处理条件下的酶活性

项目	过氧化氢酶 [μmol/(g·d)]	脱氢酶 [μg/(g·d)]	漆酶 [nmol/(g·min)]	蛋白酶 [mg/(g·d)]	磷酸酶 [μmol/(g·d)]
黑土_对照	50.88±0.43a	229.31±30.01a	666.88±35.43a	2.63±0.11a	21.27±1.66a
黑土_处理	51.28±0.70a	315.32±37.75b	716.50±18.30b	3.14±0.28b	24.48±1.38b

（续）

项目	过氧化氢酶 [μmol/(g·d)]	脱氢酶 [μg/(g·d)]	漆酶 [nmol/(g·min)]	蛋白酶 [mg/(g·d)]	磷酸酶 [μmol/(g·d)]
水稻土_对照	54.92±0.60bc	79.49±10.93c	317.03±5.54c	4.46±0.27c	10.61±0.27cd
水稻土_处理	55.42±0.66bc	67.10±9.35c	227.85±14.90d	5.32±0.15d	8.91±0.20de
绿洲土_对照	56.64±0.21c	43.98±3.36c	192.26±8.73d	6.61±0.14e	8.05±0.15e
绿洲土_处理	52.60±3.41ab	38.40±1.03c	85.97±11.68e	6.00±0.25f	7.94±0.87e
潮土_对照	54.89±0.39bc	45.80±2.67c	102.42±13.93e	3.60±0.09b	10.04±1.21cde
潮土_处理	54.41±0.28bc	58.78±5.18c	204.11±26.29d	4.32±0.26c	11.61±0.90c

注：表中数值均以"平均值±标准差"列出，同一列不同字母代表具有显著差异（$P<0.05$）。

土壤有机污染物消散的主要机制包括生物降解、挥发和光降解[173]。土壤酶活性可被认为是 PCBs 生物降解活性的指示剂。本研究中，土壤酶活性在不同土壤类型之间的变化趋势与 PCBs 消散率的变化趋势不一致。所以，挥发和光降解可能主导 PCBs 消散过程。与松散的绿洲土和潮土相比，黑土和水稻土相对紧实，土粒直径小，有机质含量高。因此，挥发和光降解过程在黑土和水稻土中受到抑制，其 PCBs 消散率低于绿洲土和潮土。有研究认为，土壤中较高的有机质含量还会降低 PCBs 的移动性[228]。

3.6 不同类型土壤中细菌菌群对 PCBs 消散的响应

通过高通量测序，为 8 个处理（$n=4$）的微生物菌群建立了 32 个 16S rRNA 基因文库，总共获得了 2 193 019 条高质量序列，平均长度 416bp。

Shannon 指数反映菌群多样性[229]。如图 3-3A 所示，对照组的菌群多样性由高到低依次为：水稻土＞潮土＞黑土＞绿洲土。处

第3章 土壤类型对PCBs消散的影响

理组的菌群多样性结果与对照组一致。Ace指数反映菌群丰富度[139]。如图3-3B所示,Ace指数与Shannon指数的结果一致,对照组和处理组的菌群丰富度由高到低依次为:水稻土＞潮土＞黑土＞绿洲土。无论对于Shannon指数还是Ace指数,每个类型的土壤在对照和处理之间都没有显著性差异(详细结果见表3-3)。从以上结果可以看出,土壤类型对菌群α多样性的影响强于PCBs污染。

文氏图(图3-3C)用于描述8个组OTU的相同点和不同点。不考虑是否染毒,4种类型土壤共享1594个OTU。与各自的对照相比,经PCBs染毒处理后,黑土的OTU保持不变(67);绿洲土的OTU从74增加到109;水稻土的OTU从66减少到53;潮土的OTU从123减少到75。因此,PCBs污染后,黑土的细菌菌群能够保持相对稳定,而其他3种土壤的菌群会受到污染的影响而发生一些变化。根据各自独有的OTU差异,可以推断出不同土壤受PCBs污染后菌群变化幅度由高到低依次为:潮土＞绿洲土＞水稻土＞黑土。

PCoA分析(图3-3D)用于描绘不同实验组菌群结构的差异和相似度,图中每个点代表1个菌群,不同颜色的点属于不同实验组,两点之间的距离越近,表明两个样本之间的细菌群落结构相似度越高,差异越小。PCoA图中的32个样本被PC1轴和PC2轴划分成3个集群。根据各点彼此之间的距离分析,绿洲土和其他土壤的菌群结构差异最大,水稻土和潮土的菌群差异最小。从PCoA图还可以发现,同一种土壤类型,无论是否受到PCBs污染,菌群都会聚集,说明土壤类型对菌群结构的影响强于PCBs污染。在4种土壤中,潮土的对照组和处理组距离最远,说明潮土受到PCBs污染后菌群结构变化最大。相反,黑土的对照组和处理组几乎完全重叠,说明PCBs污染对黑土菌群结构没有影响。PCoA分析结构与文氏分析结果一致。

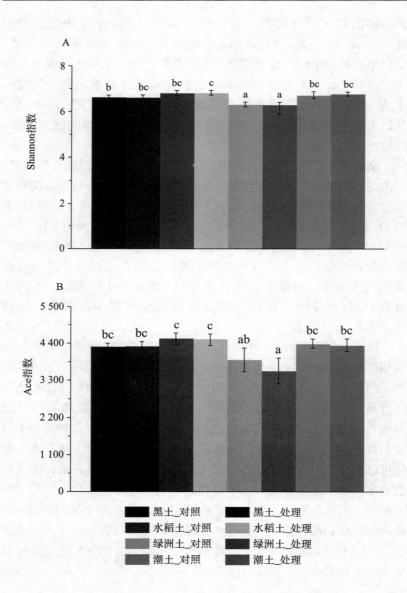

第3章 土壤类型对PCBs消散的影响

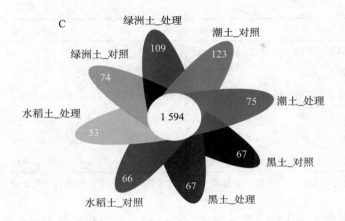

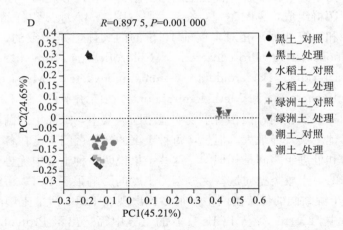

图3-3 4种土壤在不同处理条件下的细菌菌群的多样性和差异性
A. Shannon指数 B. Ace指数 C. 文氏图 D. PCoA分析

注：误差棒代表平均值±标准差，不同小写字母表示在0.05水平差异显著，$n=4$。

表3-3 4种土壤在不同处理条件下细菌菌群的α多样性

处理	Shannon指数	Ace指数
黑土_对照	6.62±0.03b	4 289.63±105.18bc
黑土_处理	6.62±0.02bc	4 312.21±135.56bc

(续)

处理	Shannon 指数	Ace 指数
水稻土_对照	6.79±0.05bc	4 530.90±169.80c
水稻土_处理	6.80±0.06c	4 511.18±177.78c
绿洲土_对照	6.29±0.09a	3 899.72±349.59ab
绿洲土_处理	6.28±0.13a	3 563.40±387.46a
潮土_对照	6.73±0.10bc	4 388.91±146.49bc
潮土_处理	6.74±0.08bc	4 353.69±174.98bc

注：表中数值均以"平均值±标准差"列出，同一列不同字母代表具有显著差异（$P<0.05$）。

为了进一步了解4种土壤的菌群结构，在门和属两个水平上研究了菌群组成，见图3-4。在门水平（图3-4A），8组样品（4组对照和4组处理）相对丰度前10位的门大体上是一致的，包括：Actinobacteria、Proteobacteria、Chloroflexi、Firmicutes、Acidobacteriota、Bacteroidota、Gemmatimonadetes、Myxococcota、Desulfobacterota 和 Methylomirabilota。在门水平，4种土壤的优势菌种虽然相同，但相对丰度明显不同。其中，绿洲土的菌群组成与其他土壤差异最大。但是，每种土壤的优势菌门丰度在各自的对照组和处理组之间是相似的。这些结果说明4种土壤中优势菌门是相似的，土壤类型对这些优势菌门丰度的影响强于PCBs污染。在4种土壤共同的优势菌门中，很多曾出现在PCBs污染土壤的研究报道中。Correa等从PCBs暴露的土壤中筛选出了Proteobacteria 和 Acidobacteria[191]。Hiraishi[230]认为，脱氯细菌主要归属于Chloroflexi、Firmicutes 和 Proteobacteria。有的研究还发现，PCBs污染后土壤中Actinobacteria 和 Bacteroidota 数量增多[197]。在本研究中，如图3-4和表3-4所示，水稻土中的脱氯菌群（Firmicutes 和 Proteobacteria）在PCBs污染后丰富度显著升高（$P<0.05$）。在4种土壤中，其他已知的脱氯细菌在门水平的丰度没有表现出显著性差异。

第3章 土壤类型对PCBs消散的影响

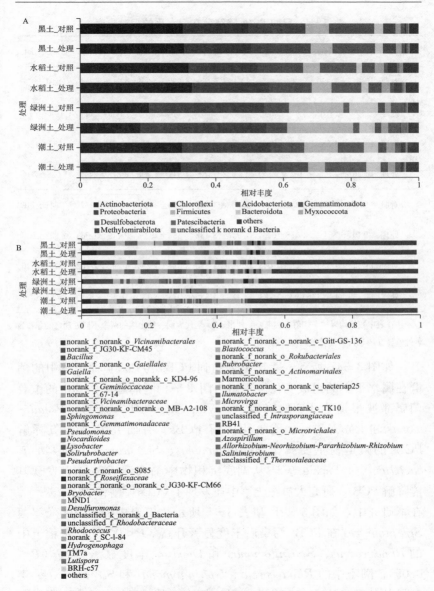

图3-4 不同处理组的细菌菌群在门和属水平的组成分析

注：相对丰度低于1%归为"others"。

表 3-4 已知 PCBs 降解菌在门水平的相对丰度

项目		Actinobacteria	Chloroflexi	Proteobacteria	Acidobacteria	Firmicutes	Bacteroidota
黑土	对照	15 065±1 390	8 364±497	9 646±1 469	6 632±1 355	3 446±316	1 291±136
	处理	14 604±398	8 454±586	9 405±338	6 005±1189	3 084±171	1 062±16
水稻土	对照	15 440±1 383	6 609±670	9 731±949	5 221±1 157	2 313±145	757±128
	处理	16 267±1 535	6 102±1 119	11 222±872	4 341±1 798	2 878±244*	1 036±50*
绿洲土	对照	9 469±782	3 383±521	15 903±1 648	859±296	7 430±367	3 941±715
	处理	9 018±551	3 078±668	18 718±3 432	883±593	9 779±1 663	2 298±958
潮土	对照	14 334±1 333	4 859±1 044	12 411±1 436	3 627±1 294	4 752±2 162	2 017±546
	处理	14 729±1 573	6 093±1 704	12 420±1 555	6 075±1 130	2 467±579	2 056±312

注：表中数值均以"平均值±标准差"列出，"*"代表同一种土壤的两个不同处理之间存在显著性差异（$P<0.05$，t 检验）。

如图 3-4B 所示，无论是否受到 PCBs 污染，4 种土壤中的菌群在属水平上的组成变化呈现出与门水平相同的趋势。其中的主要菌属都被报道为 PCBs 降解菌。Van Aken 等[194]认为，*Pseudomonas* 和 *Rhodococcus* 能氧化降解 PCBs。有研究发现需氧降解 PCBs 的微生物主要属于 *Sphingomonas*、*Streptomyces* 和 *Paenibacillus*[231]。Papale 等[232]从湖底沉积物中筛选到 *Devosia*，发现其能降解 PCBs。研究人员在土壤中还发现了 PCBs 降解菌 *Bacillus*[233]。在本研究中，如图 3-4B 和表 3-5 所示，黑土中的 PCBs 降解菌 *Streptomyces* 在 PCBs 污染后丰度显著升高（$P<0.05$）；水稻土中的 *Rhodococcus*、*Sphingomonas* 和 *Devosia* 丰度显著升高（$P<0.05$）；潮土中的 *Rhodococcus*、*Sphingomonas* 和 *Streptomyces* 丰度显著升高（$P<0.05$）。在 4 种土壤中，其他已知的 PCBs 降解菌在属水平的丰度没有表现出显著性差异。

表 3-5 已知 PCBs 降解菌在属水平的相对丰度

项目		*Pseudomonas*	*Rhodococcus*	*Sphingomonas*	*Streptomyces*	*Paenibacillus*	*Devosia*	*Bacillus*
黑土	对照	7±6	91±23	1 646±316	236±39	126±18	91±14	632±79
	处理	4±1	99±17	1 632±66	327±48*	123±11	89±6	562±27
水稻土	对照	119±128	127±43	497±58	467±54	310±16	47±11	1 187±114
	处理	256±16	216±54*	635±60*	461±72	287±21	100±25*	1 266±149
绿洲土	对照	1 646±558	51±19	262±25	124±15	271±52	331±61	2 899±86
	处理	2 267±260	197±79*	507±71*	152±11*	193±47	360±167	2 505±434
潮土	对照	439±127	304±208	698±83	189±52	216±174	61±23	1 163±681
	处理	345±45	516±99	957±152	233±93	136±10	96±38	813±172

注：表中数值均以"平均值±标准差"列出，"*"代表同一种土壤的两个不同处理之间存在显著性差异（$P<0.05$，t检验）。

土壤中的主要菌属不一致，可能是构成 PCBs 消散率差异的原因之一。有研究认为，土壤菌群差异与土壤不同的特性有关[183]。在本研究中，虽然菌群丰度不一致，但 4 种土壤中的优势菌群是一致的。但是，Chen 等[234]认为，不是优势菌群而是稀有菌群才是土壤生态功能的主要驱动者。

3.7 小结

土壤类型对 PCBs 消散、土壤酶活性和菌群起到决定性的影响。土壤 pH、阳离子交换量（CEC）、有机质含量（OM）和土粒直径（MD）对 PCBs 消散过程的影响显著。土壤类型对土壤酶活性（蛋白酶、磷酸酶、氧化还原酶、脱氢酶和漆酶）和菌群多样性

(α多样性、菌群结构和组成)的影响明显强于 PCBs 污染的影响。4 种土壤中，PCBs 污染对潮土的细菌菌群影响最明显，黑土的菌群几乎不被 PCBs 污染影响。

第4章 垃圾改良土壤对孔雀草及其根际微生物的毒性影响

4.1 引言

随着社会和经济的发展，垃圾堆积问题日益凸显，如何处理这些垃圾成为一个严峻的问题。目前，在发展中国家最有效的固废处理选项是填埋。在我国，超过80%的城市固体垃圾不经任何前处理就存放在填埋场。但是，大部分的填埋场都已达到其设计容量。所以，探寻合理有效的可循环利用填埋场垃圾的方法是非常必要的[235]。经过10年的分解腐化，矿化垃圾变得十分稳定。垃圾稳定和矿化后就可以作为一种潜在的资源被利用。同时，腾出的空间又将用于存放新的垃圾，这样就延长了填埋场的使用期限[236]。

研究人员已开始关注矿化垃圾的再利用，不仅可用作覆盖材料和建筑材料，还可用作植物生长的营养土壤。最近有研究发现，矿化垃圾可以用于石油污染土壤的生物修复[237]。因为矿化垃圾中含有丰富的溶解性有机碳化合物和营养成分，所以矿化垃圾有可能用作生草土[238]。有研究报道过一些植物可以耐受多种重金属污染，并可生长在被重金属污染的土壤，例如 *Festuca arundinacea*、*Impatiens walleriana* 和 *Tagetes patura*[239-241]。*T. patura* 是菊科的一种草本植物，广泛分布于世界各地，同时对多种重金属具有很好的耐性[242-244]。

作为土壤生态系统中重要的一部分，根际微生物参与维持植物生长和土壤污染物修复。许多研究证明，根际微生物在土壤金属矿化和植物耐受重金属等方面发挥着重要的作用[245-248]。也有研究发现，有机污染物和富营养化可以显著改变根际微生物菌群的结构[249,250]。矿化垃圾的微生物菌群可以降解污染物和降低污染物的毒性[251]。矿化垃圾的物质特征及其丰富的微生物菌群使其可能成为一种潜在的绿色资源，用于土壤和水体修复[237,243,244,252]。利用矿化垃圾作为栽培土壤或者用作改良贫瘠土壤的改良剂都具有可行性[238,253]。

本研究通过调查垃圾污染物对根际微生物和植物健康的影响来探求循环利用矿化垃圾的途径，最终建立植物修复垃圾填埋土地的基础，对于维持健康的城市生态环境具有重要的意义。

4.2 盆栽实验方案与设计

4.2.1 矿化垃圾前处理

矿化垃圾收集于山西省太原市的新沟垃圾填埋场，这个填埋场已投入使用14年，最大容量为$3.5 \times 10^6 m^3$。首先，去除垃圾中的石头、玻璃瓶和塑料等不可降解的物质。其次，将垃圾与普通表层土（收集自填埋场周边）以不同比例混合。混合比例以"普通土：垃圾"依次为：A，10∶0；B，8∶2；C，6∶4；D，5∶5；E，4∶6；F，2∶8；G，0∶10。

4.2.2 矿化垃圾和普通土壤的理化特征分析

样品的基本理化特征参考相关标准和文献进行测定。常规理化特征（pH、CEC、OM、TN和TP）测定方法见第2章。电导率（Electrical conductivity，EC）、有效氮（AN）的测定参考LY/T 1228—2015；有效磷（AP）的测定参考LY/T 1232—2015。金属元素（Cr、Pb、Cd、Cu、Zn和Mn）的测定参考HJ 803—2016。

4.2.3 植物处理方法

孔雀草种子先经3% H_2O_2浸泡30min后用纯净水反复清洗，再置于铺有双层湿纱布的托盘上进行发芽（25℃，相对湿度80%，避光)。种子发芽后转移到装有洁净土的育苗盘，放在人工气候箱进行培养（25℃±2℃）。孔雀草幼苗长出4～6片真叶后，转移到装有垃圾-土混合物的花盆进行盆栽试验。盆栽条件：25℃，65%田间持水量，90d。盆栽试验结束后，利用抖根法收集土壤样品（$n=4$），并收集植物样品（$n=5$），用于后续分析。

4.2.4 植物指标测定方法

按照试剂盒（ELISA，MLBIO company）的说明测定相关指标（$n=5$），包括：丙二醛（malondialdehyde，MDA）、蛋白羰基（protein carbonyl，PCO）、超氧化物歧化酶（superoxide dismutase，SOD）、过氧化氢酶（catalase，CAT）和过氧化物酶（peroxidase，POD）。

4.2.5 细菌菌群分析方法

主要步骤包括：DNA提取、PCR反应、高通量测序、序列加工处理和生物信息学分析。使用BLAST并比对Greengenes数据库对基因序列进行系统发育分析，置信范围为80%。最终将基因序列系统分类到门（phylum）、纲（class）、目（order）、科（family）和属（genus）5个水平。通过QIIME获得每个样品的α多样性指数（Chao1、Simpson、Shannon、Pielou_e、Observed_species、Faith_pd和Goods_coverage）。利用R Project软件进行层序聚类分析。样品间的菌群结构变化通过UniFrac距离度量，并通过主坐标（PCoA）分析呈现。基于系统分类学丰度确定每组样品的指示性菌属[254]。

4.2.6 数据分析

所用统计分析均采用单因素方差分析（ANOVA，Origin 7.0）进行。α多样性分析的数据通过四分位数范围呈现。多组样品间的α多样性差异通过Kruskal-Wallis test（with dunn'test）进行分析。

在本章所有图表中，A～G代表不同的混合比例，以"普通土∶垃圾"的比例记录，A，10∶0；B，8∶2；C，6∶4；D，

5∶5；E，4∶6；F，2∶8；G，0∶10。

4.3 矿化垃圾和普通土壤的特性分析

矿化垃圾和普通土壤的理化性质见表4-1。矿化垃圾EC值是普通土壤的1.76倍，TN含量是3.57倍，TP含量是2.13倍，Pb含量是2.27倍，Cd含量是2倍，Zn含量是2.79倍，Hg含量是2.5倍。另外，普通土壤中的Cr和As含量分别是矿化垃圾的1.42倍和1.56倍。这些结果说明垃圾已稳定和完全矿化。

表4-1 矿化垃圾和普通土壤的特性

项目	pH	EC (mS/cm)	Pb (mg/L)	Cr (mg/L)	Cd (mg/L)	Zn (mg/L)	As (mg/L)	Hg (mg/L)	TN (mg/L)	TP (mg/L)	AN (mg/L)	AP (mg/L)
矿化垃圾	7.68	3.25	38.48	46.36	0.42	163.83	8.45	0.15	2.96	1.11	176.00	18.77
普通土壤	8.50	1.85	16.92	65.89	0.21	58.67	13.18	0.06	0.83	0.52	91.93	4.86

注：表中数值为平均值（$n=3$）。

4.4 植物指标的测试结果

掺入矿化垃圾处理后，除8∶2组外，孔雀草叶片的叶绿素含量呈增加的趋势（图4-1A），与普通土壤相比，8∶2组的叶绿素含量略有下降，6∶4组、5∶5组和4∶6组的叶绿素含量显著增加。掺入矿化垃圾处理后，SOD活性逐渐增加（图4-1B），与普通土壤相比，6∶4组、5∶5组、4∶6组、2∶8组和0∶10组的SOD活性显著增加（$P<0.001$）。掺入矿化垃圾处理后，CAT活性因垃圾掺入比例的不同而不同（图4-1C），与普通土壤相比，8∶2组、5∶5组、4∶6组和2∶8组的CAT活性显著增加（$P<0.001$）。掺入矿化垃圾处理后，8∶2组和6∶4组的POD活性保

持不变，其他组的 POD 活性增加（$P<0.001$，图 4-1D）。掺入矿化垃圾处理后，MDA 含量显著下降（$P<0.001$，图 4-1E）。PCO 含量与 MDA 含量的变化趋势一致（图 4-1F）。

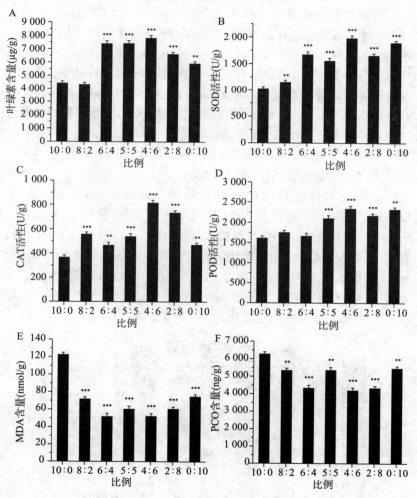

图 4-1 各组处理的植物指标

注：柱形图中的误差棒代表平均值和标准差（$n=5$）；"**"和"***"分别代表在 0.01 和 0.001 水平（ANOVA 检验）差异显著。

4.5 菌群分析结果

经扩增后,所有样品共获得 56 070 OTU,归属于 34 个门和 363 个科。各处理组在门水平的组成如图 4-2A 所示,Proteobacteria、Actinobacteria、Firmicutes、Acidobacteria、Bacteroidetes、Chloroflexi 和 Patescibacteria 是主要菌群。Firmicutes 的相对丰度在各处理组之间波动最大,在 6∶4、5∶5 和 4∶6 组丰度较高,但在 10∶0、8∶2、2∶8 和 0∶10 处理组丰度较低。各优势菌门在各组之间差异很大。各处理组丰度前 20 科的比例只占 25%~40%(图 4-2B)。与菌门分布一致,Lactobacillaceae(属于 Firmicutes)在 6∶4、5∶5 和 4∶6 组中的相对丰度高于其他各组。

多重 α 多样性指数比较分析结果,如图 4-3 所示。除了 Goods_coverage 指数以外,不同处理组的差异极显著($P<0.001$)。总体而言,矿化垃圾掺入量少(10∶0、8∶2 和 6∶4 组),菌群 α 多样性高;矿化垃圾比例高,菌群 α 多样性低。与 5∶5 和 4∶6 组相比,2∶8 和 0∶10 组的 Simpson、Shannon 和 Pielou_e 指数更高,Chao1 和 Observed species 没有显著性差异。在一定程度上,当矿化垃圾比例逐渐升高,结果正好相反。

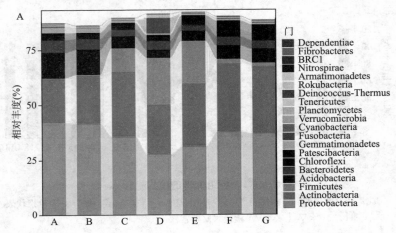

第4章 垃圾改良土壤对孔雀草及其根际微生物的毒性影响

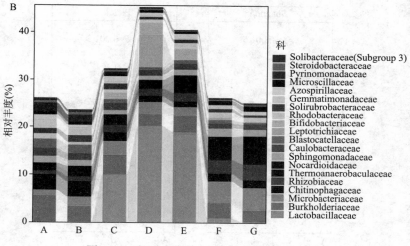

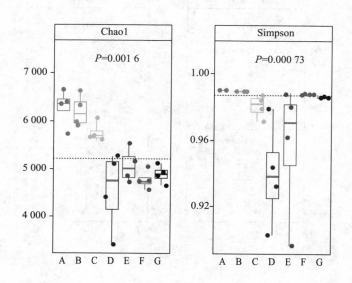

图 4-2 相对丰度前 20 的门（A）和科（B）

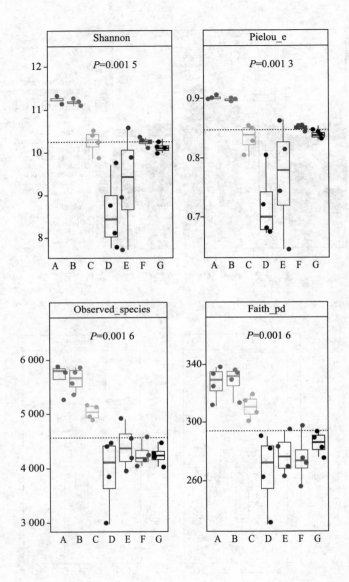

第4章 垃圾改良土壤对孔雀草及其根际微生物的毒性影响

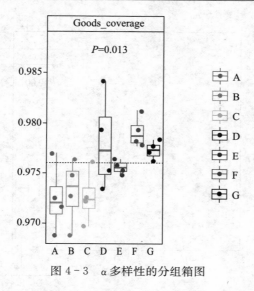

图 4-3 α 多样性的分组箱图

基于菌属丰度，对加权和未加权 Unifrac 距离矩阵进行分析，其 PCoA 分析结果如图 4-4 所示。从 10∶0 到 0∶10 组，每组的样品依次聚集分布。组间距离远大于组内距离，说明垃圾和土壤不

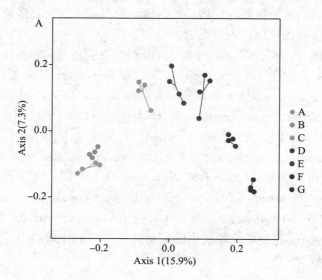

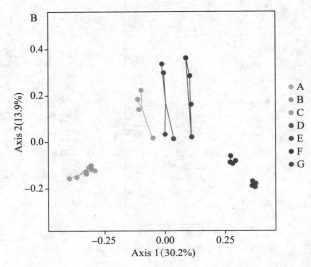

图 4-4 各种处理的菌群 PCoA 分析
A. 未加权 B. 加权的 Unifrac 距离矩阵分析图（菌属水平）

同的混合比例对根际微生物的影响不同。同时说明，掺入矿化垃圾后，主要的微生物菌群结构可能与原土的根际微生物紧密关联，关联程度又受到垃圾掺入量影响。

如图 4-5 所示，10∶0 和 0∶10 组含有明显差别的根际菌群结构。例如，原土中的 *Ferruginibacter*、*Hymenobacter*、unclassified_*Gemmataceae*、*Longimicrobium*、*Tychonema CCAP 1459-11B*、*Gemmatirosa*、uncultivated soil bacterium clone *S111* 和 *Rubellimicrobium* 丰度较高。unclassified_*Hyphomicrobiaceae*、*Polycyclovorans*、unclassified_*Phycisphaerales*、wastewater metagenome、uncultured *Crater Lake bacterium CL*120-133、*Chloroflexus*、candidate division TM7 bacterium LY2、*Oceanotoga*、unclassified_*Anaerolineae*、unclassified_*Ilumatobacteraceae* 和 *Dyadobacterare* 是纯矿化垃圾中的指示菌属。其他处理组中的指示菌属相对较少。

第4章 垃圾改良土壤对孔雀草及其根际微生物的毒性影响

图4-5 各处理组指示菌属（Genus）的指标值分析

如图 4-6A 所示，通过菌群功能预测分析，所有样品的功能丰度最高的是"生物合成"的基本途径，其次是"前体代谢物和能量产生"的途径。PCoA 分析结果显示，无论是原土比例高还是垃圾比例高，均呈现出更稳定的菌群功能模型（图 4-6B），一些中间比例的混合土菌群功能组成变化明显，不太稳定。

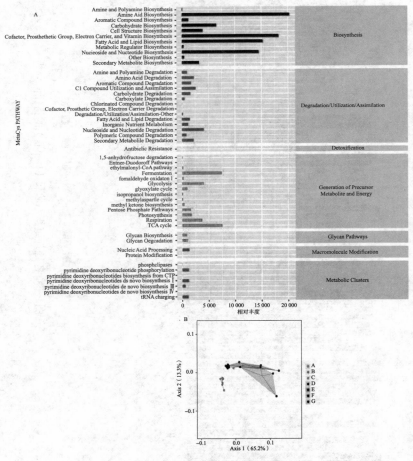

图 4-6 菌群功能预测分析

A. MetaCyc 途径丰度预测分析结果　B. 基于功能丰度的 Bray-Curtis 距离矩阵的 PCoA 分析结果

4.6 讨论

本研究的结果显示，矿化垃圾对植物生理和根际微生物都具有深远的影响。矿化垃圾激活了植物抗氧化系统和金属耐性机制，而且引发根际微生物通过调整其状态而降低污染物伤害。通过比较原土和矿化垃圾的物理与化学性质，发现矿化垃圾中含有丰富的植物生长发育所需的营养元素，例如，N、P 和 K。同时，过高的 EC 值和植物非必需的金属元素都对植物具有明显的毒害作用。这些结果与过去很多关于重金属的植物毒害作用的研究结果相一致[255-259]。活性氧（ROS）类物质反应活性高、毒性强，可引起细胞的氧化损伤，导致脂质过氧化损伤、导致膜损伤和氧化压力[260-262]。叶绿素在植物光合作用中起到重要的作用。为了保护细胞和组织免于损伤和功能紊乱，很多抗氧化酶活性提高，例如，SOD、CAT 和 POD，这些都是植物体内主要的 ROS 清除酶，用于保护膜系统[263]。重金属高污染条件下，CAT 和 POD 被激发用于清除 H_2O_2。先前很多研究都认为，在氧化压力下，SOD、CAT 和 POD 活性被激发[256,264]。

经矿化垃圾处理后，孔雀草叶部的 MDA 和 PCO 含量显著降低。矿化垃圾的掺入比例越高，MDA 和 PCO 含量越低。很明显，普通土处理组的植物体叶部 MDA 和 PCO 含量高于掺入垃圾的处理组。也就是说，在一定程度上，MDA 和 PCO 含量与叶绿素含量呈负相关关系，与抗氧化酶活性的变化相吻合。

与近期很多研究结果一致，矿化垃圾中微生物 α 多样性比普通土低[265-267]。这也说明，经过长期的生存条件筛选和微生物适应过程，矿化垃圾样品中已经形成了特有的且与周边普通土不同的微生物菌群结构。矿化垃圾处理组也有一些相对独特的菌群。很多在垃圾处理组中丰度高的菌属在木质素类废物和有机污染物的降解中起着重要的作用。*Acidobacteria bacterium SCN 69-37* 的水解酶类可以降解植物细胞壁[268]。这可能与矿化垃圾的早期成分中有木质废物有关。*Polyclovorans* 被报道与有机污染物降解有关[269,270]。*Dyadobacter* 与土壤氮代谢紧密联系，在促进植物生长和土壤修复方面起着重要作用[271,272]。*Tumebacillus* 被报道与污染物降解联系紧密，尤其是降解土壤中的邻苯二甲酸酯[273]。*Anaerosalibacter* 在石油烃降解过程中发挥关键的作用[274]。*Emticicia* 与除草剂的降解有关[275]。以上结果显示，具有不同特殊功能的很多菌种形成了一个污染物无害化的菌群，并在矿化垃圾腐熟和演替过程中发挥着关键的作用。目前，腐熟的矿化垃圾在控制固废和水体污染物方面已有一些应用。

在普通土中掺入矿化垃圾，一些原土中的优势菌群逐渐消失，例如，*Ferruginibacter*、*Hymenobacter*、*Longimicrobium*、*Tychonema CCAP 1459-11B*、*Gemmatirosa*、*Rubellimicrobium*、*Flavobacterium* 和 *Stenotrophobacter* 等。这些菌属作为土著微生物系统中主要的组成，它们在物质代谢和能量代谢、物质循环过程中发挥重要的作用。同时说明，随着有机或无机污染物的逐步增加，土壤系统的代谢功能也在发生明显的变化。这与本研究关于菌群功能预测的结果相吻合。但由于功能预测分析方法的限制，不同处理组功能代谢的差异目前只能表现在宏观尺度上。为了更深入的机理研究，有必要引入更精确和特定的组学测定策略，例如，基因组学、转录组学、代谢组学。

不同的细菌菌群逐渐适应不同的垃圾掺入比例，对降低矿化垃圾的毒性和充分利用矿化垃圾中的营养元素具有重要的作用。本研究的菌群分类学丰度和菌群功能丰度的分析结果支持这一结论[276]。

4.7 小结

与普通土壤相比，矿化垃圾中含有更丰富的营养元素，例如，有效元素（N、P 和 K）、必需金属元素（Mg、Mn、Zn 和 Cr）、非必需元素（Cd、Pb、Hg 和 As）。这些过量的金属元素对植物系

第4章 垃圾改良土壤对孔雀草及其根际微生物的毒性影响

统有不利的影响。在普通土中掺入矿化垃圾影响孔雀草叶部的叶绿素含量、抗氧化酶活性、脂质过氧化和蛋白质氧化水平,以及根际微生物菌群。无论是对植物的影响还是对菌群的影响,其程度都是依赖于掺入矿化垃圾的比例。从本研究可以看出,普通土壤掺入矿化垃圾后可能转化为更具植物栽培性的土壤。在以后的研究中还需要找出最优掺入比例,并揭示其中更深入的机制。

第 5 章 氧化石墨烯对植物根部健康的影响研究

5.1 引言

近几年，氧化石墨烯（Graphene oxide，GO）在农业领域、环保领域和材料领域等得到了广泛研究和应用[172,276,277]。GO 应用领域和应用产品的迅猛增长必将导致高水平的环境排放[147]。GO 给生态环境所带来的影响需要在其大量释放之前得以调查研究[278]。研究人员主要关注 GO 的植物毒性[279]。研究发现，GO 对植物的毒性，包括：抑制植物生长，减少叶绿素含量，引起氧化胁迫和基因毒性[280,281]。但是，GO 对植物细胞和内生菌群生态的影响仍然研究不足。

植物内生细菌是一类定殖于植物内部组织，而对植物宿主本身没有明显负作用的生物[282]。前人研究中所分离到的内生细菌主要来自植物的根部、种子和叶部，主要定殖于宿主植物的细胞间隙和维管束。内生细菌在植物细胞内定殖的报道很少[283]。研究发现，植物内生细菌能通过溶磷作用、生物固氮作用、分泌铁载体和植物激素等促进植物生长[284,285]。而且，内生细菌对植物宿主还有很多其他的有利影响，例如，抑制病原菌、提高植物抗逆能力和生物修复作用[286,287]。水稻是一种主粮作物[288]，主要通过根部摄取营养和水分，所以，水稻根部的内生细菌对水稻的生长、健康和生物量起着决定性作用。作为外来物质，GO 对水稻根部内生菌群的影响不能被忽视。但是，因为对于植物内生细菌对生长条件的要求还不完全知晓，所以，通过常规的培养技术对植物内生微生物多样性进行分析是不足的[288]。

在本研究中,水稻幼苗在水培条件下暴露于 5mg/L GO 培养 15d。研究 GO 暴露对水稻根部形貌、细胞亚显微结构和内生细菌菌群的影响,利用高通量测序技术对内生细菌菌群的变化进行了研究。

5.2 水培实验方案与设计

本研究所用 GO 购自南京先丰纳米材料有限公司,纯度:>99%;粒径分布:342~532nm;C:71.23%;O:28.77%;厚度 0.8~1.4nm。本实验其他特用试剂见表 5-1,特用仪器见表 5-2。

表 5-1 实验试剂

试剂名称	生产厂家	纯度
丙二醛(MDA)测定试剂盒(TBA 法)	南京建成生物工程研究所	—
过氧化物酶(POD)试剂盒	南京建成生物工程研究所	
BCA 法蛋白含量测定试剂盒	南京建成生物工程研究所	
过氧化氢酶(CAT)试剂盒	南京建成生物工程研究所	
超氧化物歧化酶(SOD)试剂盒(WST-8 法)	南京建成生物工程研究所	
LB 培养基	南京建成生物工程研究所	—
次氯酸钠	天津市福晨化学试剂厂	分析纯
70%乙醇	天津市福晨化学试剂厂	分析纯

表 5-2 实验仪器

名称	型号	厂商
电子天平	Sartorius BSA124S	赛多利斯科学仪器有限公司
冷冻超薄切片机	Leica EM FC7	德国徕卡
光学显微镜	Olympus ZL 61	Olympus,Tokyo
DXR 智能拉曼光谱仪	Thermo Scientific DXR2xi	Thermo Scientific

5.2.1 水培实验过程

催芽：水稻（*Oryza sativa* L.）种子在 5% H_2O_2 中浸泡 30min，用去离子水彻底冲洗干净。双层纱布铺在培养托盘（30cm×40cm）中，用超纯水（18.2Ω/cm）润湿纱布。将 150 粒形状和大小基本一致的种子撒在纱布上，用保鲜膜盖在托盘上，托盘放在培养箱中进行催芽。催芽条件：温度 25℃，相对湿度 80%，光照 3 000lx。

GO 悬浊液的配制：0.25mg GO 加入 50mL 超纯水中，超声 10min（功率 400W），得到 5mg/L 的 GO 悬浊液。

水培试验：催芽 1 周后结束，获得水稻幼苗。将 5 株幼苗转移到装有 GO 悬浊液的 50mL 离心管中，用海绵球固定幼苗，如图 5-1 所示，同时做 20 个平行处理（$n=20$）。然后，将离心管中的幼苗置于光照培养箱中进行培养，培养条件：光照 3 000lx，温度 25℃，相对湿度 80%。为了减少 GO 沉淀，每天 3 次轻微地涡旋搅拌（150r/min）GO 悬浊液，每次 10min。其间，每天用超纯水补充蒸发和损失的水分，维持离心管中的液面在 40~50mL。同时，将催芽后的水稻幼苗移植到装有纯水的离心管中进行培养作为对照（$n=20$）。处理组和对照组均按照相同的步骤培养 15d 后，水培结束。

图 5-1 水培试验部分样品照片

共设置 2 组处理，每组处理 20 个重复，"Control"代表没有 GO 添加的对照处理；"GO"代表有 GO（5mg/L）暴露的处理。

5.2.2 根部形貌、细胞结构和氧化压力分析方法

水稻根部经无菌水彻底清洗后，利用光学显微镜观察根部形貌，利用透射电镜观察细胞亚显微结构。利用金刚石刀片制备 80nm 根部切片。利用拉曼光谱仪分析根部 GO 的成分，拉曼光谱仪在 780nm 条件下激发光谱，激光的光斑为直径 $1\mu m$ 的圆，激光照射位置为根部横切面。按照 Song 等[289]的方法分析检测根部抗氧化酶（SOD、POD 和 CAT）的活性和丙二醛（MDA）的含量。以上研究均做 3 个重复，$n=3$。

5.2.3 根表面消毒方法

在参考相关文献[288]的基础上稍作修改来进行根表面消毒，从先到后依次经过以下步骤：无菌水彻底冲洗；浸入 70％乙醇保持 3min；搅动条件下用次氯酸钠溶液（2.5％ Cl^-）冲洗 5min；浸入 70％乙醇保持 30s；无菌水冲洗 5 次。根部消毒最后一遍的冲洗水利用细菌培养方法验证根部是否彻底消毒。100mL 清洗水无菌移至 LB 细菌培养基平板，30℃培养 72h 观察是否有菌落生长。如果没有菌落生长，说明根部消毒有效，无菌的根样品进行后续分析。

5.2.4 植物内生菌菌群分析方法

1. DNA 提取和 PCR 反应 称取 0.5g 消毒后的新鲜水稻根（一式三份）进行 DNA 提取，按照 Fast DNA SPIN extraction kits（MP Biomedicals，Santa Ana，CA，USA）试剂盒的说明进行操作。通过 PCR 反应扩增细菌 16S rRNA V5～V7 区的 DNA 片段。每个样品要进行 3 次技术重复。

引物（primer）：799F（5′-AACMGGATTAGATACCCKG-

3′)、1193R（5′-ACGTCATCCCCACCTTCC-3′）；这对引物能够将植物叶绿体或线粒体的基因最大程度排除掉[290]。

2. 高通量测序和生物信息学分析 原始序列已经上传至 NCBI Sequence Read Archive（SRA）数据库（Accession Number：PRJNA506642）。

利用 MOTHUR 获得每个样品的稀释曲线（rarefaction curves）、菌群丰度指数 Chao 和菌群多样性指数 Shannon。使用 RDP Classifier 并结合 silva（SSU115）16S rRNA database 对 16S rRNA 基因序列进行系统发育分析，置信范围为 70%[291]。利用 R Project 软件进行层序聚类分析。通过绘制文氏图（Venn）描述样品间的相似性和差异性[229]。菌群分类信息通过 BugBase 数据库进行菌群表型特征预测。

5.2.5 数据分析

所有实验都进行 3 次重复，所得的数据均进行方差分析，文中误差棒代表了标准差，结果以 SPSS（19.0）统计软件进行分析，数据的显著水平均指 $P<0.05$（ANOVA，t-test）。

5.3 植物根部生理和生化变化

如图 5-2 所示，经 GO 暴露的水稻根尖颜色比对照更黑，可能是 GO 在根尖的大量积累所致。与对照相比，暴露组根表面出现了裂口，根尖的根毛也消失了。GO 的纳米片层结构具有十分锋利的边缘[292]，水稻根部在与 GO 的交互作用过程中失去了完整性。水稻根表面出现的裂口有助于根部对 GO 的提取和积累。不仅如此，GO 还可能通过渗透作用进入植物根的内部[293]。以前有研究也发现，小麦在石墨烯暴露条件下根毛被破坏[294]。

如图 5-3 所示，通过 TEM 观察根细胞的亚显微结构发现，GO 暴露组的根细胞轮廓变得不规则。对照组根细胞展现出完整的细胞壁、细胞膜、细胞核以及其他的细胞结构。暴露组根细胞出现

第 5 章 氧化石墨烯对植物根部健康的影响研究

图 5-2 光学显微镜下水稻根尖的变化

了明显的质壁分离和 GO 沉积。据报道，GO 通过细胞膜的自发渗透和 GO 增强的膜渗透进入细胞膜，沉积在细胞内[295,296]。

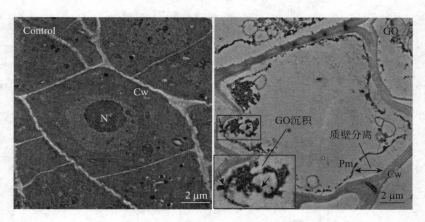

图 5-3 透射电子显微镜（TEM）下的根细胞
注：Cw，细胞壁；Pm，细胞膜；N，细胞核。

随后通过拉曼光谱分析进一步鉴定了根的横切面，如图 5-4 所示，暴露组根截面谱图出现了 GO 的典型 D 峰和 G 峰，对照组没有出现 GO 的特征峰。以上结果说明 GO 确实进入了根内部。

GO 暴露对水稻根生物化学参数的影响，见图 5-5。GO 暴露后，SOD 和 POD 活性显著升高，MDA 含量也显著升高，说明 GO 暴露对植物根造成氧化胁迫和细胞损伤。

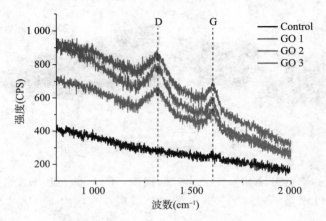

图 5-4　根截面的拉曼光谱分析图

注：GO1～3 代表 3 个不同 GO 处理的根样品谱图。

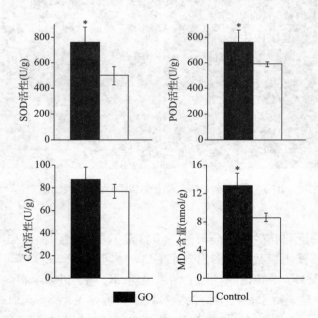

图 5-5　GO 暴露对根的氧化胁迫

注：误差棒代表平均值±标准差（$n=3$）；* 表示在 $P<0.05$ 水平具有显著性（t 检验）。

5.4 根部内生细菌菌群丰富度和多样性分析

水稻根经表面消毒后,选用植物内生菌特用引物 799F 和 1193R 进行 DNA 提取,然后经 PCR 扩增和高通量测序技术测序,两组样品共获得超过 21 000 个高质量序列。基于 97% 的相似性,暴露组和对照组分别平均获得 57 个和 72 个 OTU,见表 5-3。OTU 和 Chao 指数反映微生物菌群的种群丰富度,所以,GO 暴露后水稻根部内生细菌菌群丰富度降低。Shannon 指数反映种群多样性[229],GO 暴露后 Shannon 指数从 2.51 降低到 1.72,说明内生细菌菌群多样性降低。与对照组相比,暴露组的 Shannon 指数低于对照组,而 Simpson 指数高于对照组,说明 GO 暴露后内生细菌菌群的均匀度下降[297]。作为具有强化学活性的外来物质,GO 的出现必将引起供内生菌定殖的根内部微生态环境的变化,内生菌菌群丰富度和多样性也将随之发生改变。

表 5-3 水稻根部内生细菌菌群的丰富度和多样性

处理	Reads	3% distance			
		OTU	Chao	Shannon	Simpson
GO	21 960±230a	57±5a	61±3a	1.72±0.02a	0.297 2±0.033a
Control	27 601±498b	72±6b	72±5b	2.51±0.15b	0.161 2±0.027b

注:误差棒代表平均值±标准差($n=3$)。

5.5 根部内生细菌菌群结构和表型分析

基于 97% 的相似性,内生菌分类信息如图 5-6 所示。在门水平上,优势菌群包括:Proteobacteria、Firmicutes、Actinobacteria、Bacteroidetes 和 Fusobacteria。在纲水平上,优势菌群包括:Gammaproteobacteria、Bacilli 和 Alphaproteobacteria。在属水平上,暴露组和对照组的菌群共发现 10 个菌属,其中 5 个菌属是一致

的，包括：*Cronobacter*、*Lactococcus*、*Pseudomonas*、*Rhizobium* 和 *Yersinia*，共有的菌群分别占暴露组和对照组各有菌群的 54.10% 和 56.35%。"Unclassified"代表不能被划归为已知分类组的序列，可能是仍未被收录的菌属或者宿主植物细胞器的基因信息[298]。

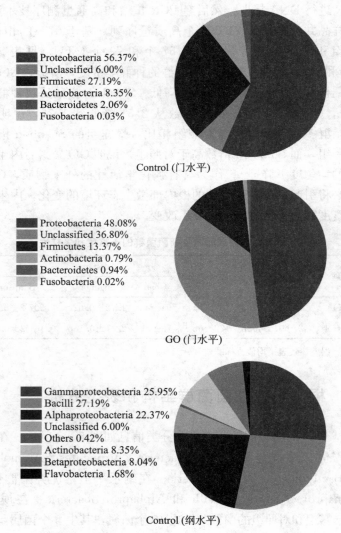

Control (门水平)

GO (门水平)

Control (纲水平)

第5章 氧化石墨烯对植物根部健康的影响研究

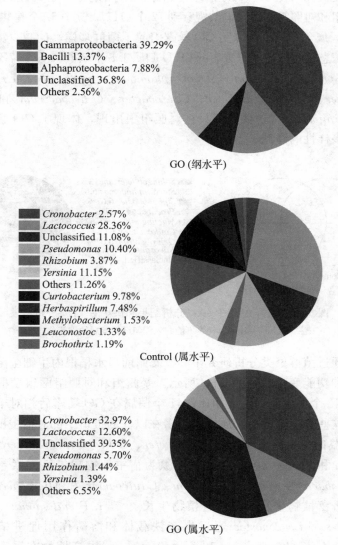

图 5-6 内生细菌菌群在门、纲和属水平的分类信息

注:"Others"代表在纲和属水平上相对丰度<1%的菌群。

文氏图（图5-7A）描绘暴露组和对照组菌群的共性和差异。暴露组和对照组两组菌群共观察到72个OTU，其中57个是共有的，15个只属于对照组，说明GO暴露引起了菌群多样性的减少。通过分类信息比对，对照组独有的OTU归属于8个菌群（图5-7B），包括：*Sphingobacterium*、*Stenotrophomonas*、*Siphonobacter*、*Procabacter*、*Mucilaginibacter*、*Caulobacter*、*Ochrobactrum*和*Novosphingobium*。这些菌群只在对照组中出现，体现了GO暴露后菌群多样性的降低。

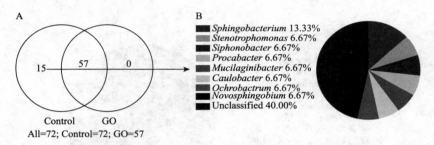

图5-7 菌群文氏图（A）分析结果和饼状图（B）分析结果

注："Unclassified"代表不能归入任何已知菌属的细菌菌群。

通过菌群聚类分析研究了GO暴露前后水稻根内生细菌菌群的结构和功能变化。如图5-8所示，暴露组和对照组两组菌群共观察到43个菌属，与对照相比，41个菌属在GO暴露后相对丰度下降。这些丰度降低的菌属中有很多在以往的报道中都对宿主植物具有有益的影响。例如，*Acinetobacter*、*Ochrobactrum*、*Herbaspirillum*和*Rhizobium*通过生物固氮为宿主植物提供氮元素[283,299]；*Arthrobacter*、*Methylobacterium*、*Caulobacter*和*Microbacterium*通过分泌植物生长激素促进植物生长[300,301]；*Pseudomonas*、*Bacteroides*和*Leuconostoc*通过生产铁载体和溶磷作用促进植物生长[302-304]；*Sphingomonas*和*Curtobacterium*提高植物在重金属污染土壤的生长能力[305,306]；*Pseudomonas*、*Bacillus*和*Paenibacillus*通过抑制病原菌影响植物健康[307-309]。因此，GO暴露后，这些

第5章 氧化石墨烯对植物根部健康的影响研究

内生菌群丰度下降将会影响水稻的生长和健康。

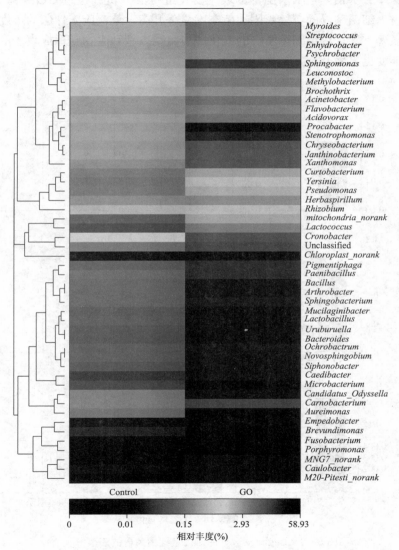

图5-8 暴露组和对照组菌群在属水平的聚类分析

如图 5-9 所示，基于 16S rRNA 基因序列，通过 BugBase 数据库对水稻内生细菌菌群的表型变化进行了分析。经过 GO 暴露，内生细菌菌群的革兰氏阴性菌丰度显著升高，革兰氏阳性菌的丰度显著下降；内生细菌菌群的氧化胁迫耐受表型和生物膜形成表型的丰度也显著升高。

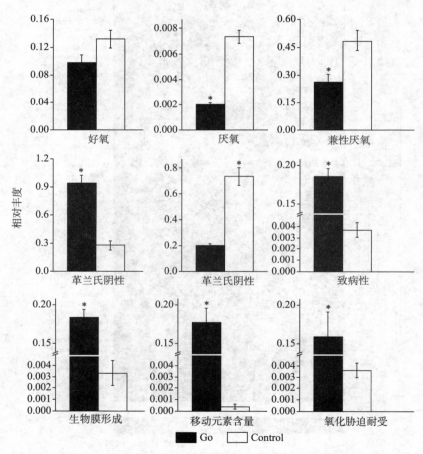

图 5-9 BugBase 数据库对内生细菌菌群的表型分类

第5章 氧化石墨烯对植物根部健康的影响研究

GO进入水稻内部也将与内生菌接触，GO能引发超氧自由基的生成，细胞膜的破坏以及导致细胞物质的流出[310,311]。GO在与内生菌接触的过程中，其锋利的边缘会引起细胞膜破坏。革兰氏阴性菌的细胞壁比阳性菌的细胞壁多一层，更能抵抗细胞膜的损伤[312]。GO引发的氧化压力导致内生菌群氧化胁迫耐受表型丰度的升高。内生菌细胞物质从损伤的细胞流出，促进了生物膜形成表型丰度的升高，生物膜的形成有利于细胞耐受不利的环境压力[313]。GO暴露引起了根内微生态环境的变化，内生菌群表型的改变使其适应能力增强。

5.6 小结

水培条件下，GO暴露导致水稻根部表面损伤、根尖细胞结构破坏和根内GO沉积。GO进入水稻根内，导致内生菌群丰富度、均匀度和多样性的降低，导致很多对水稻宿主有益的菌群丰度下降，导致菌群的表型特点改变，例如，革兰氏阴性表型、氧化胁迫耐受表型和生物膜形成表型的丰度升高。但是，GO的植物毒性和GO对内生菌群的影响之间的关系还不清楚，需要进一步研究。

第 6 章 植物根系分泌物对氧化石墨烯特性的影响

6.1 引言

近年来，氧化石墨烯（Graphene oxide，GO）在多个领域得到广泛研究和应用，例如：生物、化学、医药和污染治理[134,314,315]。相关报道认为，GO 及其衍生产品的市场在 2020 年达到 6.75 亿美元[316]。由于 GO 产品的大量生产和应用，其环境排放量也将大大增加[147]。正因如此，很多研究开始关注 GO 的植物毒性，例如，GO 可以抑制植物生长，破坏细胞结构，降低叶绿素含量，诱导氧化应激和基因毒性[317-320]。虽然 GO 厚度接近 1nm，但它在自然环境中并不是惰性的。在评估其环境风险之前，应该全面了解环境因子和自然有机质对 GO 的修饰作用[321-323]。研究结果证明，在可见光照的水中，石墨烯类材料的形貌由纳米片状转变成带状。有研究发现，自然界中的有机质可以以配体的形式自发地固定在石墨烯表面，致使其环境归属和环境风险改变[324,325]。最近的研究报道也证明，微生物可以通过电子转移改变氧化石墨烯的表面化学性质[326-330]。植物体是自然生态系统中的初级生产者，其对 GO 在形貌、结构、表面化学和相应环境风险的影响仍然不得而知[147,331-333]。

作为植物的地下"隐藏"部分，植物根系的基本功能通常被认为是固定植物体和促进营养、水分的摄取。不过，根系同样可以向周围的环境分泌大量种类繁多的根系分泌物[334]。根系分泌行为被认为是根系与环境之间一种主要的非常复杂的交互作用，例如植物

第 6 章 植物根系分泌物对氧化石墨烯特性的影响

-污染物交互作用[335-337]。研究发现，在重金属暴露的环境压迫下，作为防御机制，植物根系会分泌电解质类、糖类、有机酸类、氨基酸类、酶类、激素类和次级代谢产物以适应环境[338]。目前，对于根系分泌物如何影响重金属和其他传统污染物的生物可利用性有了一定的了解，但是，根系分泌物与包括 GO 在内的纳米材料的交互作用却不得而知。与其他纳米材料相比，GO 具有特殊的纳米片层形貌、含氧官能团，其更容易结合小分子物质[147]。因此，一旦根系分泌物作为自然配体固定在 GO 表面，其特性和环境风险也会随之改变[339]。

为了验证上述设想，本研究以水稻（*Oryza sativa* L.）作为一种单子叶模式植物，通过 GO 刺激其根系分泌行为，然后检测根系分泌物种类。这些根系分泌物作为配体（ligands）与原始氧化石墨烯（PGO）结合，形成复合物质（LGO）。

随后，本文对比研究了 PGO 和 LGO 的物理化学特性，包括：形貌、结构、表面官能团、粒径分布、光吸附特性和表面电荷（孤对电子）等。最后，以斑马鱼作为模式动物，通过研究 PGO 和 LGO 对斑马鱼的孵化率、存活率、生长发育和线粒体功能的影响，初步分析了根系分泌物作为纳米材料的配体对环境的影响。

6.2 水培实验方案与设计

本实验所特用试剂见表 6-1，本实验所特用仪器见表 6-2。

表 6-1 实验试剂

试剂名称	生产厂家	纯度
乙酸乙酯	天津市康科德科技有限公司	色谱纯
氧化石墨烯（GO）	南京先丰纳米材料有限公司	纯度＞99%
N-甲基-N-（三甲基硅烷基）三氟乙酰胺（MSTFA）	百灵威科技有限公司	纯度＞98.5%
N, O-双（三甲基硅烷基）三氟甲基乙酰胺（BSTFA）	百灵威科技有限公司	纯度＞98.5%

表6-2 实验仪器

名称	型号	厂商
场发射能量过滤透射电子显微镜（TEM）	JEM-2010FEF	JEOL，Japan
多功能扫描探针显微镜（AFM）	Nanoscope 4	Veeco，USA
ZETA电位+广角激光散射仪（绿光532nm）	ZETAPALS/BI-200SM	Brookhaven，NY，USA
X射线能谱仪（XPS）	Axis Ultra DLD	Kratos Analytical Ltd.
傅立叶变换红外光谱仪（FTIR）	Tensor 27	Bruker，Germany
拉曼光谱仪（Raman）	Renishaw inVia	Renishaw plc，UK
荧光显微镜	Olympus ZL 61	Olympus，Tokyo

6.2.1 水培实验步骤

1. 水培试验和样品 LGO 收集　如图6-1，水培试验方法和步骤同第5章。PGO悬浊液的配制：5mg PGO加入50mL超纯水中，超声10min（功率400W），得到100mg/L的PGO悬浊液。

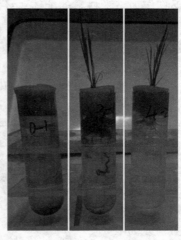

图6-1　水培试验部分样品照片

第6章 植物根系分泌物对氧化石墨烯特性的影响

LGO 收集：水培结束后，用 $0.1\mu m$ 滤膜对离心管中的悬浊液进行抽滤，将 LGO 抽滤到滤膜上，同时获得抽滤液水相，然后在 $-55℃$ 下对滞留在滤膜上的 LGO 进行冷冻干燥获得粉末状样品。

2. 分离和鉴定吸附在 LGO 表面的根系分泌物

（1）称取 0.01g LGO 粉末置于 10mL 离心管中，并向其中加入 1mL 提取溶剂（体积比，甲醇：氯仿：超纯水＝2.5：1：1），进行微波辅助萃取（40℃，15min），然后离心 10min（11 000g），收集上清液。同时，按照同样的步骤对沉淀进行二次萃取，并合并两次萃取得到的上清液。

（2）转移上清液到干净的 10mL 离心管中，加入 $500\mu L$ 超纯水，离心 3min（5 000g）后，溶液分层，上层溶液是甲醇/水相，下层为氯仿相。甲醇/水相冷冻干燥，用氮气吹干氯仿相。

（3）衍生化。首先加入 $50\mu L$ 用吡啶溶解的 BSTFA（20mg/mL），密封，涡旋，离心 5min（4 000g），30℃ 温浴 90min。而后加入 $80\mu L$ 硅烷化试剂 N-甲基-N-（三甲基硅烷）-三氟乙酰胺（MSTFA），37℃ 温浴 30min。转移衍生化好的样品到适合 GC-MS 分析的内衬管中。

（4）GC-MS/MS 分析。

气相色谱进样参数：进样量 $1\mu L$，进样口温度 230℃，不分流进样模式，载气为氦气，流速 2mL/min，使用自动进样器进样。对于高浓度样品，分流进样模式，分流比设定为 1：25。

气相色谱参数：MDN-35 毛细管色谱柱（30m），温度程序为 80℃ 恒温 2min，然后以 15℃/min 的速率升温到 330℃，然后持续 6min。传输线温度设定为 250℃。

质谱参数：离子源温度设定为 250℃，质荷比：70～600，采集速率每秒 20 次，质谱电子轰击源灯丝开启时间在色谱溶剂延迟 170s 后，检测器电压 1 700～1 850V，质谱亏损设置为 0，灯丝偏置电流为 70V，仪器自动调谐。

谱图解卷积参数：力可公司自带的商业软件 Chroma TOF，基线消除（baseline offset）设置为 1；谱图平滑（smoothing）为 5

数据点，峰宽（peak width）3s；信噪比（signal-tonoise ratio，S/N）为10。鉴定色谱库：NIST 08 library。

3. 分离和鉴定水培抽滤液中的根系分泌物　通过液液萃取法，向抽滤后的水相中加入乙酸乙酯萃取根系分泌物，得到乙酸乙酯相为中性组分；将剩余水相用1mol/L HCl调节pH=2后，再用乙酸乙酯萃取，得到乙酸乙酯相为酸性组分；将剩余水相再用1mol/L NaOH调节pH=8，用乙酸乙酯萃取，得到乙酸乙酯相为碱性组分。将酸性、碱性和中性乙酸乙酯萃取液分别旋蒸（45℃）浓缩至1mL，然后按照上述方法进行衍生化并分析鉴定。

4. 纳米材料的表征方法

（1）形貌和结构表征。样品粉末分散于无水乙醇中，超声10min，进行扫描电子显微镜（SEM）、透色电子显微镜（TEM）和原子力显微镜（AFM）分析。

（2）zeta电位和粒径分布分析。将样品粉末分散于pH 5～9的超纯水中（1.0mg/L），上机ZETAPALS/BI-200SM分析。激光器：30mW、635nm固体激光器。

（3）表面化学官能团分析。利用X-射线光电光谱仪（XPS）和傅立叶变换红外光谱仪（FTIR）对样品进行检测分析。FTIR分析条件：分辨率2cm^{-1}；扫描波段400～4 000cm^{-1}。XPS分析条件：全谱扫描的能量范围为0～1 200eV；扫描步长为1eV；分析面积为0.7mm×0.3mm；分辨率为0.2eV；分析软件：Casa-XPS V2.3.13。

（4）拉曼光谱分析（Raman）。扫描范围100～4 000cm^{-1}；分辨率：0.5 cm^{-1}；紫外可见光谱分析（UV-vis）：扫描范围180～800cm^{-1}；分析软件：UVWin5。

（5）电子顺磁共振（EPR）分析。X-频段9.55GHz；磁场调频100kHz；2，6，6-tetramethyl-1-piperidinyloxy（TEMPO）作为自由基猝灭剂。0.01mg/L的样品水溶液加入石英玻璃管中上机检测。

5. 斑马鱼孵育实验　孵化96d的斑马鱼胚胎移入盛装E3培养

基（5mmol/L NaCl、0.17mmol/L KCl、0.33mmol/L CaCl$_2$、0.33mmol/L MgSO$_4$，pH 7.4）的 24 孔细胞培养板中，每板 20 个胚胎，3 个重复。其间，用显微镜观察并记录斑马鱼胚胎的孵化率和畸变率。畸形主要包括与正常发育状态不同的仔鱼形态，例如，骨骼畸形、心包囊肿、脊柱弯曲等。

6. 线粒体膜电位分析　用 JC-1 荧光染料对斑马鱼仔鱼染色后，用 E3 培养基进行冲洗后在荧光显微镜下观察。分析软件：CellSens Standard 1.6 software。

6.2.2　数据分析

所有实验都进行三次重复，如有特殊情况，会在文中图表标注说明。文中误差棒代表了标准差，结果以 SPSS 19 统计软件进行分析。使用单因子变异数和图基（Tukey）事后检验法分析比较数据，统计显著性设为 $P<0.05$。根系分泌物通过软件 MultiExperiment Viewer 4.8.1. 进行聚类分析。

本章结果中，"PGO"和"LGO"分别代表原始氧化石墨烯和经根系分泌物修饰的氧化石墨烯。

6.3　根系分泌物鉴定结果

利用 GC-MS/MS 分析，每个样品对 200 个色谱峰进行鉴定，共鉴定了 62 种根系分泌物，见图 6-2、表 6-3、表 6-4 和表 6-5。这些根系分泌物包括醇类、酚类、萜类、有机酸类、糖类、氨基酸类、醚类、醛类、烯烃类、烷烃类和其他小分子物质。在没有 PGO 暴露的对照组，检测到 25 种根系分泌物；在暴露组的水相和吸附相，分别检测到 31 种和 47 种分泌物。如图 6-2 所示，与对照组相比，暴露组中的丙氨酸（Alanine）、十八烯酸（Octadecenoic acid）、乙醇（Ethanol）、戊酸（Pentanoic acid）、辛醇（Octanol）、十八炔酸（Octadecynoic acid）、Ethanimidic acid、油醇（Oleyl alcohol）、异樟醇（Isoborneol）和香叶基香叶醇（Gera-

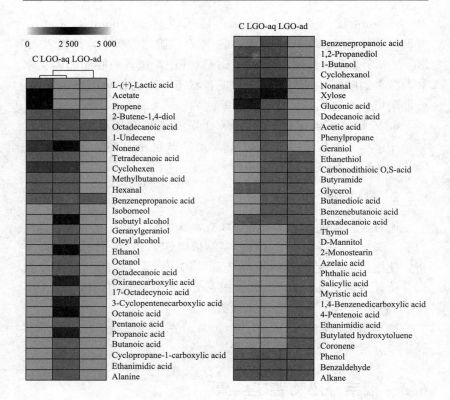

图6-2 热图分析植物根系分泌物的相对含量

注:"C"表示没有氧化石墨烯暴露的对照组根系分泌物;"LGO-aq"表示暴露组的水相中根系分泌物;"LGO-ad"表示暴露组氧化石墨烯上吸附的根系分泌物;详细数据见表6-3、表6-4和表6-5。

nylgeraniol)的含量增加。正如植物的防御机制,在PGO的外部刺激下,根系分泌物增多[340]。与暴露组水相中的分泌物相比,LGO吸附相中的百里香酚(Thymol)、D-甘露糖醇(D-mannitol)、2-甘油硬脂酸酯(2-monostearin)、壬二酸(Azelaic acid)、邻苯二甲酸(Phthalic acid)、水杨酸(Salicylic acid)、肉豆蔻酸(Myristic acid)、苯二羧酸(1,4-benzenedicarboxylic acid)、晕苯(Coronene)和草酸(Ethanedioic acid)的含量升高。在暴露组中,

第6章 植物根系分泌物对氧化石墨烯特性的影响

根系分泌物在水相和吸附相的分布由分泌物和 PGO 两者的化学性质决定[341]。在吸附相中，有机酸分子含有一个或多个羧基，次级代谢产物含有苯环。而 PGO 分子含有多种含氧官能团，例如，—OH、—COOH 和—CHO，以及多种特殊的纳米结构，例如，纳米孔、高活性边缘、悬空碳和 sp^2 结构[186,342]。上文提到的根系分泌物的化学官能团能通过氢键和 π-π 键与 PGO 分子结合。这些数据说明 PGO 暴露刺激根部分泌根系分泌物，导致 PGO 与这些小分子物质结合并形成 LGO。最近，有研究指出 PGO 可以通过静电作用吸附溶菌酶[343]和通过平面效应吸附苯类物质[344]。因为根系分泌物种类繁多，所以上述所讨论的物理和化学作用都有可能参与 PGO 与根系分泌物之间的相互作用。

表 6-3 无 PGO 暴露的对照组根系分泌物

序号	名称	相对丰度
1	Cyclohexanol	113 002±3 251
2	Glycerol	207 376±10 775
3	1，2-Propanediol	6 344±169
4	1-Butanol	12 662±605
5	Phenol	19 692±157
6	2-Butene-1，4-diol	23 810±1 238
7	Methylbutanoic acid	19 795±675
8	Acetic acid	41 467±2 126
9	Gluconic acid	4 139±313
10	Hexadecanoic acid	291 474±10 537
11	Octadecanoic acid	313 936±6 923
12	Benzenepropanoic acid	26 716±1 062
13	Dodecanoic acid	1 049±64
14	tetradecanoic acid	71 505±2 454
15	L-（＋）-Lactic acid	17 753±726
16	Acetate	2 993±89

(续)

序号	名称	相对丰度
17	Xylose	1 239±28
18	1-Undecene	118 451±5 444
19	Cyclohexen	4 098±86
20	Nonene	4 160±82
21	Propene	2 345±141
22	Hexanal	4 680±271
23	Benzaldehyde	18 111±343
24	Phenylpropane	6 714±286
25	Alkane（C2、C3、C5、C8、C10、C12、C13、C15、C16、C18）	345 187±11 450

表 6-4　PGO暴露组的水相中氧化石墨烯

序号	名称	相对丰度
1	Cyclohexanol	185 344±11 034
2	Glycerol	240 874±4 642
3	1，2-Propanediol	12 974±1 057
4	1-Butanol	28 568±853
5	Phenol	10 726±513
6	2-Butene-1，4-diol	18 087±667
7	Isoborneol	5 623±253
8	Isobutyl alcohol	3 032±167
9	Geraniol	12 834±431
10	Geranylgeraniol	9 809±517
11	Oleyl alcohol	4 811±537
12	Ethanol	2 287±123
13	2-Octanol	9 706±162
14	Ethanethiol	22 256±1 206
15	Methylbutanoic acid	23 367±742

第6章 植物根系分泌物对氧化石墨烯特性的影响

(续)

序号	名称	相对丰度
16	Acetic acid	151 031±5 379
17	Gluconic acid	19 947±1 193
18	Hexadecanoic acid	264 882±6 745
19	Octadecanoic acid	289 647±5 612
20	Benzenepropanoic acid	52 321±1 124
21	Dodecanoic acid	5 360±321
22	Tetradecanoic acid	71 505±2 375
23	Octadecenoic acid	4 403±130
24	Oxiranecarboxylic acid	3 336±76
25	17-Octadecynoic acid	17 463±141
26	3-Cyclopentenecarboxylic acid	3 641±73
27	Octanoic acid	2 332±69
28	Pentanoic acid	4 845±120
29	Propanoic acid	3 089±79
30	Butanoic acid	1 023±83
31	Butanedioic acid	15 155±654
32	Cyclopropane-1-carboxylic acid	8 357±165
33	Benzenebutanoic acid	14 696±890
34	Ethanimidic acid	70 222±3 587
35	Carbonodithioic O, S-acid	203 914±7 342
36	Butyramide	50 639±2 349
37	Alanine	6 383±532
38	Xylose	3 573±268
39	1-Undecene	101 669±863
40	Cyclohexen	4 098±432
41	Nonene	2 285±159
42	Hexanal	5 586±321

(续)

序号	名称	相对丰度
43	Nonanal	1 427±83
44	Benzaldehyde	16 816±843
45	Benzenesulfonic acid	8 512±456
46	Phenylpropane	28 007±1 403
47	Alkane (C2、C3、C5、C8、C10、C12、C13、C15、C16、C18)	80 141±8 756

表 6-5　PGO 暴露组中 LGO 吸附相的根系分泌物

序号	名称	相对丰度
1	Glycerol	384 050±7 296
2	Phenol	109 727±5 596
3	Ethanethiol	22 256±178
4	Bisphenol	49 636±1 346
5	Thymol	1 600 696±85 836
6	D-Mannitol	473 459±19 411
7	Geraniol	16 994±611
8	Hexadecanoic acid	2 415 764±125 619
9	Octadecanoic acid	211 500±6 345
10	2-Monostearin	151 424±4 537
11	Azelaic acid	110 322±3 971
12	Benzenepropanoic acid	25 315±699
13	Phthalic acid	37 753±1 515
14	Benzenebutanoic acid	51 395±2 021
15	1,4-Benzenedicarboxylic acid	140 847±3 239
16	Salicylic acid	365 692±13 164
17	Ethanimidic acid	70 222±1 631
18	4-Pentenoic acid	20 444±1 034

(续)

序号	名称	相对丰度
19	Myristic acid	55 058±1 385
20	Carbonodithioic O, S-acid	203 914±10 191
21	Ethanedioic acid	154 734±2 321
22	Butanedioic acid	86 591±1 771
23	Butylated hydroxytoluene	125 458±6 398
24	Undecene	22 543±202
25	Coronene	2 455 149±61 471
26	Butyramide	50 639±208
27	Propiophenone	25 572±251
28	Benzaldehyde	25 081±332
29	9-Acetylphenanthrene	78 996±157
30	Cyclohexene	1 823 776±9 299
31	Pentacosane	33 140±1 230

6.4 根系分泌物引发氧化石墨烯形貌的改变

纳米材料的形貌与它的环境归宿和纳米毒性都联系紧密[147,345,346]。如图6-3所示的SEM图，PGO和LGO都具有纳米片层形貌，但LGO比PGO质地粗糙。在图6-3的TEM结果中，由于根系分泌物的修饰，LGO纳米片层的透明度比PGO明显降低。在图6-4的AFM结果中，PGO纳米片的厚度为0.862nm，符合单层氧化石墨烯0.8~1.0nm的标准[347]；LGO纳米片在厚的区域为3.856nm，在薄的区域为1.591nm，平均厚度大致为2.265nm。纳米片由于固定了根系分泌物而厚度增加。由于在AFM检测的前处理过程中，需要长时间的超声处理，所以AFM图片中的纳米片尺寸差异较大。

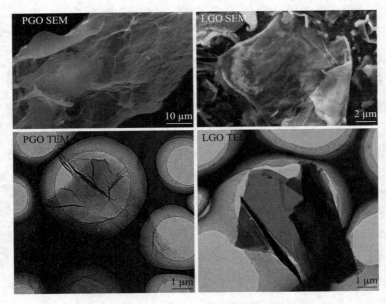

图 6-3　PGO 和 LGO 的扫描电子显微镜（SEM）和
透色电子显微镜（TEM）表征结果

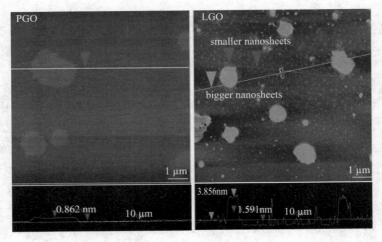

图 6-4　PGO 和 LGO 的原子力显微镜（AFM）表征结果

6.5 根系分泌物引发氧化石墨烯粒径分布和表面电荷的改变

纳米材料的粒径分布和表面电荷对其环境行为和相关的生态风险起主要的作用[147,348,349]。如图 6-5 所示，PGO 粒径分布在342～450nm 之间，而 LGO 粒径分布在 142～220nm 之间，说明根系分泌物使 PGO 的尺寸减小，但引起这种尺寸分布变化的机制仍不得而知。图 6-6 所示纳米材料电位在不同 pH 条件下的 zeta，PGO 和 LGO 在自然环境所涉及的 pH（pH 5～9）范围间都具负电荷。而且，在所测试的 pH 范围内，PGO 带有更多的负电荷，比 LGO 更稳定。一般而言，生物细胞表面带负电荷。本研究的结果意味着，与 PGO 相比，LGO 通过静电相互作用将更容易与生物细胞结合[350]。LGO 表面所减少的负电荷与所固定的具有高碳氧比的根系分泌物有直接的关系。

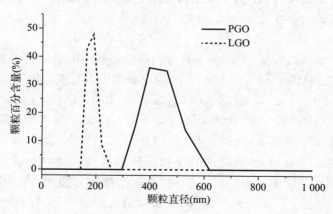

图 6-5　PGO 和 LGO 的粒径分布分析结果

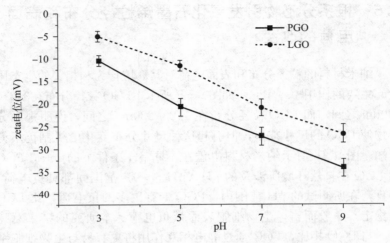

图6-6　PGO 和 LGO 的 zeta 电位分析结果

6.6　根系分泌物引发氧化石墨烯表面化学官能团的改变

　　XPS 可以表征 PGO 在根系分泌物修饰前后的表面化学官能团的变化。从图 6-7 PGO 和 LGO 的全谱图中可发现，PGO 谱图由 69.73% C1s 和 30.27% O1s 组成；而 LGO 谱图的组成包括 71.49% C1s、25.06% O1s、3.30% N1s 和 0.15% S2p。通过分析，可以发现 LGO 的碳氧比（C/O）高于 PGO。很明显，这种变化的原因是 LGO 表面的分泌物配体具有较高的碳含量，如图 6-2 所示的晕苯、壬二酸、甘油-硬脂酸酯和百里酚。如图 6-8 所示，通过 FTIR 分析，在 LGO 表面检测到亚甲基（Methylene）官能团。亚甲基的检出表明在 LGO 表面引入了碳链，这个结果也呼应了 LGO 的高碳氧比。

　　图 6-9 呈现了 XPS 精细谱的结果，PGO 的 C1s 谱由 44.98% 碳-碳信号（C—C/C=C）和 55.02% 碳-氧（C—O/C=O）信号组

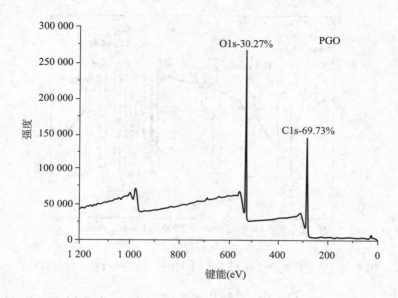

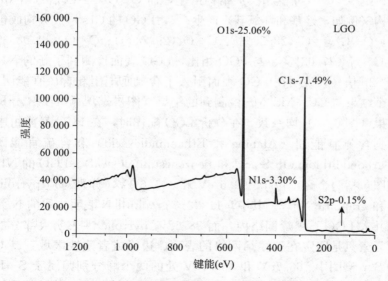

图 6-7　X 射线电子能谱（XPS）分析 PGO 和 LGO 的全扫描谱图

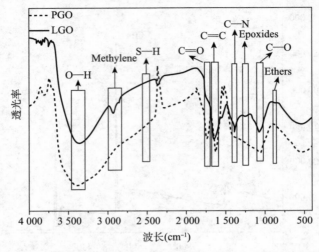

图6-8 傅里叶变换红外光谱仪（FTIR）表征 PGO 和 LGO 的结果

成。C—C/C=C 信号由 sp^2 结构产生，而 C—O/C=O 信号由 sp^3 结构（例如，羟基和环氧基）产生[351]。LGO 的 C1s 谱成分构成包括：C—C/C=C（28.25%）、C—O（29.74%）、C=O（14.98%）和 C—N（27.03%）。与 PGO 相比，LGO 表面检测到了新的 N1s 和 S2p 信号，说明在 LGO 表面引入了含氮官能团和含硫官能团，这个结果与 GC—MS/MS 检测到的结果（图6-2）和 FTIR 分析结果（图6-8）均一致。在分析 LGO 配体时，在 PGO 暴露组中测到含氮官能团（Alanine 和 Ethanimidic acid）和含硫官能团（Carbonodithioic O、S-acid 和 Benzenesulfonic acid）。LGO 的 N1s 谱图包括两个峰，位于 399.6 eV 和 401.6 eV，分别对应 pyrrolic N 和 graphitic N[352]。其中，11.32% graphitic N 表明少量的 N 元素并入 LGO 石墨烯网络内，而 88.68% pyrrolic N 表明大部分的 N 元素只是固定在石墨烯网络的表面、边缘或者残缺区域。LGO 的 S2p 谱图中 163.7eV 和 168.7eV 处的两个峰分别归属于 S—H 键和 S—O 键[353,354]。XPS 分析结果与根系分泌物配位体的检测结果相一致。同时，上述结构表明在 LGO 表面引入了含氮官能团和含

第6章 植物根系分泌物对氧化石墨烯特性的影响

硫官能团,与纳米片厚度分析和表面电荷的分析相呼应。

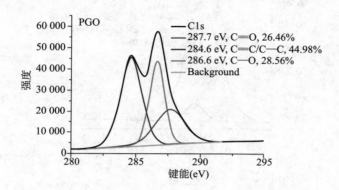

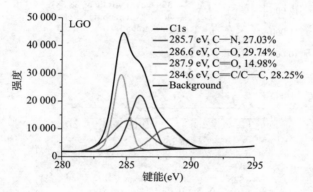

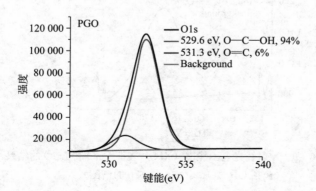

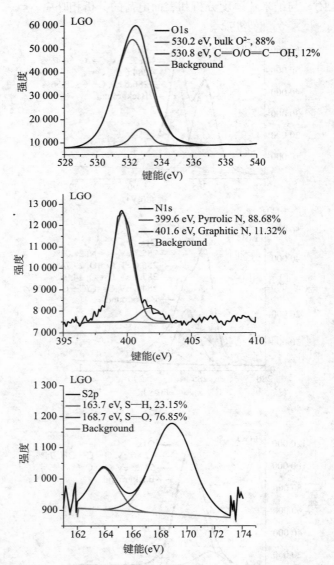

图 6-9 X 射线光电子能谱(XPS)表征 PGO 和 LGO 的结果
（包括 C1s、O1s、N1s 和 S2p 的谱图）

6.7 根系分泌物引发氧化石墨烯表面化学性质的改变

纳米材料表面的孤对电子与表面结构缺陷有关，直接影响其化学性质和纳米毒性[355-357]。以 TEMPO 作为孤对电子捕获剂，利用 ERP 进行孤对电子测试。如图 6-10 所示，PGO 和 LGO 都表现出强于对照的信号，说明两种纳米材料都能够产生孤对电子。相同浓度的分散液中，LGO 产生的孤对电子稍多于 PGO，说明 LGO 表面固定的根系分泌物配体可能增强了其化学活性。众所周知，碳纳米材料的孤对电子与其不规则的结构有关，例如，悬空键、边缘或内部平面的转角和缺陷等[147,358]。本研究通过 Raman 对 PGO 和 LGO 的结构进行分析，如图 6-11 所示。D 带（1 363cm^{-1}）和 G 带（1 593cm^{-1}）分别反映了石墨烯的无序结构和对称有序的 sp^2 碳结构[359]。D 带和 G 带的强度比值（D/G）与样品的尺寸和微晶大小有关，常用来衡量其结构无序程度，比值越大，其无序程度越高。LGO 和 SGO 的 D/G 分别为 1.86 和 1.15，说明根系分泌物作为配体增强了 PGO 结构的无规则程度。拉曼分析结果与孤对电子分析结果相吻合。紫外-可见光吸收光谱用来揭示 PGO 和 LGO 的光化学活性，如图 6-12 所示。PGO 和 LGO 谱图中的最大吸收峰都大概位于 202～204nm，用来反映 n-π* 和 π-π*（E2 带）转

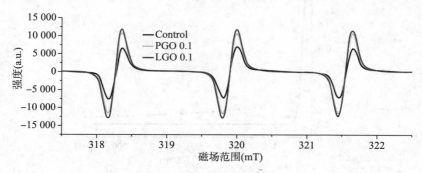

图 6-10　PGO 和 LGO 的电子顺磁共振光谱（ERP）分析结果

移[360]。但是，LGO 谱图最大吸收峰的强度明显高于 PGO 的谱图，反映了来自自然配体的—COOH、—NH$_2$—、—S—、不饱和的酮基和孤对电子存在于 LGO 表面。与 PGO 相比，LGO 在 240~800nm 范围的 UV-vis 吸收明显减少，说明在固定根系分泌物后石墨烯片层内的电子共轭被打乱[361]。与 PGO 相比，由于 LGO 高效率的 n-π*/π-π* 转移，其表现出可见光条件下吸收变弱，而在紫外光照射下吸收变强。

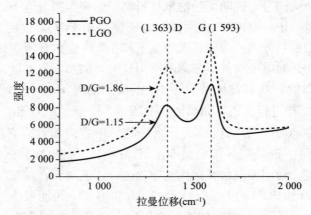

图 6-11 PGO 和 LGO 的拉曼光谱 (Raman) 分析结果

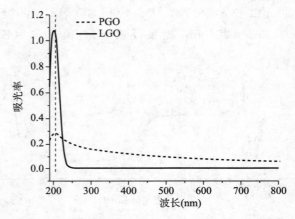

图 6-12 PGO 和 LGO 的紫外-可见光光谱 (UV-vis) 分析结果

6.8 环境影响分析

毫无疑问，材料性质的改变会引起其生态风险的改变。例如，碳纳米管在形貌、结构和表面化学性质的变化就会引起其在环境介质稳定性和生物体内氧化应激的改变[362-364]。根系分泌物配体使PGO在环境pH内的稳定性和表面携带的负电荷发生变化。为了探讨PGO固定根系分泌物后所引起的纳米毒性变化，本章以模式生物斑马鱼作为研究对象，分析PGO和LGO对其发育的影响，如图6-13所示。对照处理中，孵化率为83.3%。而PGO和LGO暴露组都引起孵化率的下降，在浓度1mg/L和10mg/L暴露72h，孵化率为63.4%~68.5%。PGO暴露组和LGO暴露组间的孵化率差异不明显。与对照相比，PGO和LGO暴露组也都没有引起明显的胚胎死亡率升高。但是，LGO暴露组引起了明显的斑马鱼发育畸形，如尾部弯曲和心包囊肿。在LGO暴露浓度为1~10mg/L时，畸形率达到4.2%~5.8%。在低浓度（0.1mg/L）暴露时，暴露组间没有明显的差异。

线粒体对纳米颗粒异常敏感，有研究发现纳米ZnO和纳米银能引起线粒体膜电位的下降[365,366]。如图6-14所示，以JC-1染料对96d的鱼仔染色后观察线粒体膜电位变化，发现PGO和LGO暴露下膜电位在下降。与对照相比，在0.1~10mg/L PGO和LGO暴露下，鱼仔绿色荧光强度显著增加，表明线粒体膜电位的下降，会引起线粒体功能的缺失。与PGO暴露组相比，LGO暴露组的红/绿荧光强度比下降了17.9%~75.5%。以上结果表明，PGO在所试浓度范围可以提高斑马鱼发育畸形率和引发膜电位下降。但是在自然环境中，PGO的浓度还是不得而知。在以往的研究中，由老化和环境因素所引起的对纳米TiO_2、纳米ZnO和碳纳米管的修饰作用主要钝化了这些纳米材料的活性，提高了它们的稳定性，进而减轻了生物体对它们的氧化应激[331,332,367]。但是，本研究发现，根系分泌物作为修饰配体降低了PGO的稳定性和提高了其化学活性。这些改变能够引发斑马鱼的发育畸形率升高,线粒体膜电位降低。

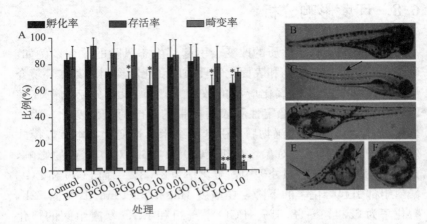

图 6-13　PGO 和 LGO 对斑马鱼孵化率、存活率和畸变率的影响

A. 柱状图描述 PGO 和 LGO 对斑马鱼发育的影响　B. 对照的斑马鱼图片,没有 PGO 和 LGO 的暴露　C. PGO 暴露下的斑马鱼鱼尾弯曲图片　D. LGO 暴露下的心包囊肿图片　E. LGO 暴露下斑马鱼鱼尾弯曲和心包囊肿　F. LGO 暴露下的斑马鱼孵化延迟

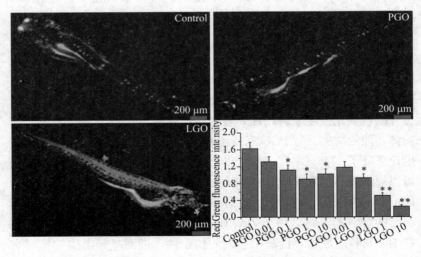

图 6-14　PGO 和 LGO 对斑马鱼线粒体膜电位的影响

注:"Red:Green fluoresscence intensity"代表红色荧光和绿色荧光强度比。

6.9 小结

通过原始氧化石墨烯（PGO）刺激植物根部分泌根系分泌物，其作为配体（Ligands）固定在 PGO 表面形成 LGO。这些根系分泌物主要由小分子有机物组成，包括有机酸类、醇类、烷烃类、氨基酸类和其他次级代谢产物。与 PGO 相比，LGO 呈现厚度增加、成分碳氧比升高、尺寸下降和透明度下降。LGO 表面出现了新的含氮和含硫官能团。LGO 比 PGO 结构更无序，具有更多的孤对电子和具有更强的紫外吸收。LGO 表面携带的负电荷低于 PGO，比 PGO 稳定性低。根系分泌物修饰 PGO 所引起的这些性质变化致使模式生物斑马鱼发育畸形率显著升高和线粒体膜电位下降。

第7章 纳米材料对小麦种子萌发能力的影响研究

7.1 引言

近年来，纳米材料在多个领域得到广泛研究和应用，例如：生物、化学、医药和污染治理等，随之，纳米材料的环境排放量也将大大增加[138]。如今正处于纳米产业发展的初期，在纳米材料大批量释放到环境之前，应该仔细评估其对生态环境的影响[139]。其中，粮食作物作为初级生产者，纳米材料的谷物毒性尤其值得关注。纳米材料是指在三维空间中至少有一维处于纳米尺度范围（1~100nm）或由它们作为基本单元构成的材料[140]。常见的纳米材料类型包括：碳基纳米材料（石墨烯、碳纳米管和富勒烯等）、金属基纳米材料（纳米金属单质或者氧化物）和复合材料等。

据报道，纳米材料对植物种子萌发能力具有促进作用还是抑制作用，仍然没有形成统一的认识[368]。Siddiqui 等[141]研究发现，纳米 SiO_2 促进番茄种子萌发和早期幼苗的生长。Daohui 等[142]研究认为，纳米 ZnO 抑制玉米种子发芽和幼苗根伸长。还有研究发现，多壁碳纳米管对萝卜种子萌发没有影响，而纳米 Ag 却使萝卜种子发芽率降低 20%，根伸长减少 70%[143]。为此，本研究通过选用不同尺寸的多种纳米材料，包括：金属基纳米材料（纳米银粉和纳米银片）和碳基材料（单壁碳纳米管、多壁碳纳米管、单层石墨烯和石墨烯纳米片），研究它们对小麦种子发芽率和根长的影响，以期揭示纳米材料对小麦种子萌发的影响。

7.2 萌发实验方案与设计

小麦种子：常规品种，购于山西太谷鑫晋农种业有限公司。
纳米材料：购于南京先丰纳米材料科技有限公司，特征见表7-1。

表7-1 纳米材料种类及其尺寸

纳米材料	尺寸特征
纳米银片	片径：～5μm；BET：0.80～1.45m^2/g
纳米银粉	粒径：80～90nm
单壁碳纳米管	直径：1～2nm；长度：1～3μm
多壁碳纳米管	直径：～50nm；长度：0.5～2μm
单层石墨烯	片径：0.5～5μm；厚度：0.8～1.2nm 单层
石墨烯纳米片	片径：～5μm；厚度：1～5nm

纳米材料溶液的制备：称取0mg、2mg、10mg、20mg、40mg纳米材料分别置于200mL盛有超纯水的广口塑料瓶中，拧紧瓶盖，超声振荡15min，分别制得0mg/L、10mg/L、50mg/L、100mg/L、200mg/L的纳米溶液，其中，0mg/L作为对照。

种子表面消毒：选取籽粒饱满、大小一致的小麦种子，将种子用3‰H_2O_2浸泡10min后，自来水冲洗3次，超纯水冲洗3次，用滤纸将水吸干。

种子发芽实验：将表面消毒的40粒种子平铺于放有双层滤纸的培养皿中，分别向培养皿中加入预先配置的不同种类和浓度的纳米材料溶液至种子高度的1/3～1/2位置。然后放入培养箱中发芽6d，培养箱温度为25℃，相对空气湿度为70%，每个处理设置3个重复。

以幼根至少达种子长度、幼芽至少达种子1/2长度作为发芽的标准，计算发芽率，并测量种子根长。将每个培养皿40粒种子的发芽率和根长求平均值，再根据3个重复求出标准差。

文中误差棒代表了标准差，结果以SPSS 20统计软件进行分析。使用单因子变异数和图基（Tukey）事后检验法分析比较数据，统计显著性设为 $P<0.05$；使用 Origin8.5 软件制图。

7.3 纳米银粉和纳米银片对小麦种子萌发的影响

如图 7-1A 所示，不同浓度的纳米银粉处理后小麦的发芽率均低于对照（56.7%），说明纳米银粉对小麦发芽具有抑制作用。暴露浓度在 50mg/L 时小麦的发芽率最低（32.5%），对小麦种子的发芽抑制作用最大，差异显著（$P<0.05$）；暴露浓度在 10mg/L、100mg/L 和 200mg/L 时，小麦的发芽率差异不显著。纳米银片的暴露浓度在 100mg/L 时发芽率高于对照，但差异不显著；其他浓度的发芽率都低于对照，差异显著（$P<0.05$），暴露浓度为 10mg/L 时小麦发芽率最低（30.0%）。总体来看，在纳米银粉和纳米银片的所试浓度范围内（10～200mg/L），小麦发芽率低于对照，说明颗粒状和片状的纳米银都对小麦发芽具有抑制作用；在所试浓度范围内，随着暴露浓度的升高，发芽率都呈现先下降后上升的总体趋势，说明纳米银在低浓度时对小麦发芽抑制作用更强。Thuesombat 等[369]的研究发现，纳米银抑制水稻种子的发芽，但随着纳米银粒径和浓度的增加，抑制作用逐渐增强。还有研究发现，纳米银颗粒可以抑制黑芥菜（*Brassica nigra*）种子的发芽。

纳米银粉和纳米银片对小麦种子根长的影响如图 7-1B 所示。不同浓度的纳米银粉处理后，发芽后小麦根长均低于对照；纳米银粉浓度 50mg/L 时根长达到最低，介于 50～200mg/L 时小麦根长差异不显著，总体来看，纳米银粉抑制小麦种子根伸长。纳米银片的暴露对小麦根伸长也多为抑制作用，纳米银片对小麦根长的影响趋势与其对发芽率的影响趋势相吻合。Pittol 等[370]的研究发现，银纳米颗粒促进萝卜根生长，却对洋葱根生长具有抑制作用。还有研究认为，纳米银对豌豆种子根伸长有促进作用[371]。这可能是由于不同品种作物的种子对纳米银的暴露响应表现不同。从本研究的数据发现，颗粒状和片状的纳米银对小麦种子萌发总体呈现抑制作用。

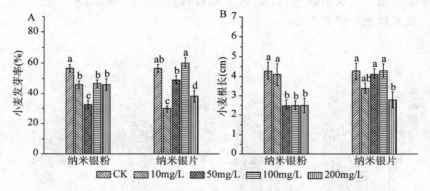

图 7-1　不同浓度的纳米银粉和纳米银片对小麦种子萌发的影响
注：不同小写字母表示在 $P<0.05$ 水平差异显著。

7.4　单壁碳纳米管和多壁碳纳米管对小麦种子萌发的影响

如图 7-2A 所示，不同浓度单壁碳纳米管（SWCNTs）和多壁碳纳米管（MWCNTs）的暴露处理下，小麦种子发芽率变化趋势一致：都在 50mg/L 处理时发芽率最低（33.4% 和 44.2%），明显低于对照（56.7%），差异显著（$P<0.05$）；都在 10mg/L 处理时略高于对照处理，差异不显著。Khodakovskaya 等[372]研究发现，碳纳米管（CNTs）能渗透到番茄种子内部进而影响种子萌发，在 10～40mg/L 暴露时能提高种子发芽率。还有研究也发现，低浓度的 CNTs 可以促进水稻种子发芽和根系生长，高浓度的 CNTs 反而产生毒害作用[373]。

SWCNTs 和 MWCNTs 对小麦种子伸长的影响趋势相似，如图 7-2B 所示。在所试浓度范围内（10～200mg/L），小麦种子在萌发阶段，根伸长几乎不受 CNTs 的影响，所有处理组根长与对照相比存在很小差异,差异均不显著。Begum 等[374]通过水培实验发现,红苋菜、生菜和黄瓜在 MWCNTs 的暴露下根长显著下降,羊豆角和大豆不敏感。Yan 等[375]的研究却发现,SWCNTs 会促进玉米种子的根系生长。

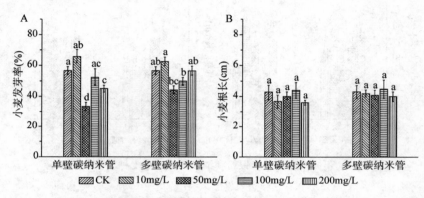

图 7-2 不同浓度的碳纳米管对小麦种子萌发的影响

本研究所使用的两种 CNTs 材料,长度接近,但直径相差很大,SWCNTs 直径为 1~2nm,MWCNTs 直径为 50nm,但两者对小麦种子萌发的影响基本一致:抑制发芽率,对根伸长作用不明显。

7.5 单层石墨烯和石墨烯纳米片对小麦种子萌发的影响

根据层数不同,将石墨烯分为:单层石墨烯,厚度 0.8~1.2nm;石墨烯纳米片,厚度 1~5nm。这两种不同尺寸的石墨烯对小麦种子发芽率的影响见图 7-3A 所示,在两种材料的暴露下,小麦种子发芽率的变化趋势一致。在所试浓度范围内(10~200mg/L),两种石墨烯材料暴露条件下,小麦种子发芽率先升高后下降。与对照相比,在单层石墨烯 10mg/L 暴露时,小麦种子发芽率(66.7%)达到最高值,高于对照(56.7%),差异显著

（$P<0.05$）；石墨烯纳米片在相同暴露浓度时，小麦种子发芽率（65.9%）也达到最高值，高于对照，差异不显著。暴露浓度超过50mg/L，两种石墨烯材料都表现为抑制小麦种子发芽。

如图7-3B所示，在所试暴露浓度范围内（10～200mg/L），单层石墨烯和石墨烯纳米片对小麦发芽种子根长的影响趋势都分别与发芽率的变化趋势一致，低浓度促进根伸长，高浓度抑制根伸长。Zhang等[376]的研究证实了石墨烯可以促进小麦根伸长。还有研究发现，低浓度的氧化石墨烯促进油菜种子的萌发，高浓度（超过25mg/L）的氧化石墨烯处理显著抑制油菜根伸长[377]。有研究还发现，经过20d的暴露，石墨烯对卷心菜、番茄、红苋菜生长的抑制作用明显，但对生菜无影响[378]。

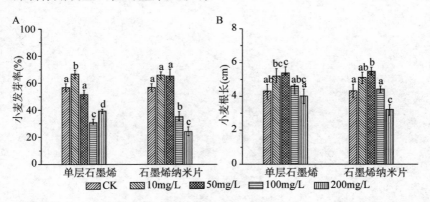

图7-3 不同浓度的单层石墨烯和石墨烯纳米片作对小麦种子萌发的影响

综合本研究中的发芽率和根长变化，发现单层石墨烯和石墨烯纳米片在低浓度时（<50mg/L）促进小麦种子萌发，在高浓度时（>100mg/L）抑制小麦种子萌发。

7.6 小结

本研究探究了不同纳米材料对小麦种子萌发的影响，并比较了相同材质不同尺寸的纳米材料对其萌发的影响。研究发现，颗粒状和片状的纳米银对小麦种子萌发总体呈现抑制作用。单壁碳纳米管和多壁碳纳米管对小麦种子萌发的影响也基本一致：抑制发芽率，对根伸长作用不明显。单层石墨烯和石墨烯纳米片均在低浓度时促进小麦种子萌发，在高浓度时抑制小麦萌发。可见，不同纳米材料对小麦种子萌发的影响不同，但不同尺寸的同种材质纳米材料对小麦种子萌发的影响无明显差别。

第8章 氧化石墨烯在土壤中性质改变和引起的土壤细菌菌群变化

8.1 引言

氧化石墨烯（Graphene oxide，GO）由于带有含氧官能团的优势，常常作为一种合成石墨烯基纳米材料的优良前体物质，应用于多个领域[314,317,321,379]。最近几年，GO 的生产量和使用量急剧升高，将必然造成其环境排放量的增加[323,380,381]。相比水和大气环境，土壤生态系统很有可能成为最大的纳米污染物接收地[382-384]。如今正处于纳米产业发展的初期，在纳米材料大批量释放到环境之前，应该仔细评估其对环境的影响[278]。原始氧化石墨烯（PGO）有可能改变土壤微生物菌群的结构和功能。同时，在 PGO 与土壤成分的交互作用中，PGO 的性质也可能发生改变。在土壤中，PGO 可以被土壤微生物修饰，也可以通过静电相互作用而吸附土壤矿物质等[385]。因此，通过研究了解 PGO 和土壤成分的交互作用以及通过交互作用形成的土壤修饰后的氧化石墨烯（SGO）非常重要，而且对于评估氧化石墨烯的环境归宿和环境风险非常关键[172]。

有研究表明富勒烯（C_{60}）和碳纳米管（CNTs）在低浓度时对土壤微生物菌群作用有限[278,386]，而在 5 000mg/kg 的高浓度暴露条件下作用显著[382,387]。但是，氧化石墨烯（PGO）对土壤微生物菌群的影响仍然不清楚。目前为止，针对 PGO 对微生物影响的研究主要集中于纯培养和水环境。研究人员在纯培养和工程水系统中发现 PGO 抑制微生物的活性[314,388-390]。有研究报道 PGO 通过与 *Escherichia coli* 的交互作用，含氧官能团的数量减少了 60%[327]，这种 PGO 被微生物还原的现象同样被真菌和 *Shewanella* 得到验证[321,328,391]。另外，有研究还报道反硝化细菌能在 PGO 表面掺杂进氮元素[330]。土壤成分极其复杂，微生物菌群种类繁多，所以，研究土壤对 PGO 的修饰作用不能仅仅在水溶液和纯培养条件下进行。微生物菌群是土壤中有环境压力存在时的一类灵敏的指示生物，但 PGO 对土壤微生物菌群结构和功能的影响研究比较缺乏[382,387,392]。而且，PGO 进入土壤环境，其对微生物菌群的改变和自身的性质改变是相伴发生的。现有的一些研究主要针对单菌，其不能真实反映自然环境对 PGO 性质的修饰改变。将土壤修饰后的 SGO 从土壤中分离出来，并对其进行表征和定性分析才能真实地调查研究 PGO 性质的真实变化。

为了验证以上的假设和弥补研究缺陷，本研究主要关注 PGO 和土壤的交互作用。本研究在 PGO 进入土壤培养 90d 后，利用高通量测序技术对土壤中的细菌菌群变化进行了调查分析。随后，本研究分析表征了 SGO 的形貌、表面化学性质和表面修饰的小分子物质。最后，通过比较 PGO 和 SGO 的粒径分布、表面电荷、孤对电子、无序结构、紫外-可见光吸收特性和氧化还原电位等性质揭示了 PGO 被土壤修饰后化学性质的改变。

8.2　土壤培育实验方案与设计

主要实验材料同第 6 章。

将 PGO 按一定的浓度比（1.0g/L）分散到去离子水中，通过超声处理使其形成稳定的悬浊液，具体的步骤详见参考文献[392]。实验用土来自天津北辰区红光农场（39°11′N、117°01′E）农田的表层土（0～20cm）。土壤风干后过 1mm 孔径的尼龙网筛以去除石头，土块等杂质。称取去除杂质的干土 1 500g 转移到塑料花盆（平均直径 18cm、高 12cm，3 个重复）中，浇水并调整到田间最大持水量的 50%。然后将花盆放在 25℃条件下预培养 7d，使土壤微生物活性稳定。培养结束后，将提前准备好的 PGO 悬浊液与花盆中的土样充分搅拌混合。为了在土培实验结束后有助于 SGO 的分离,本研究设置的 PGO 浓度较高,为 5g/kg(干土)。对照组花盆加入相同体积的去离子水。最后把所有花盆放到温室中培养(25℃)。其间,用去离子水维持土壤水分保持在田间最大持水量的 50%。土培实验进行 90d 后,从花盆的底部、中部和上部分别取样并混合,收集的土样用于土壤微生物菌群分析和 SGO 的分离与表征。

8.2.1 分离 PGO 和 SGO

土培结束后的土壤鲜样先经过冷冻干燥,然后过 0.15mm 孔径网筛,获得干燥的土壤粉末。称取 10g 土壤粉末转移到 50mL 离心管中(20 个重复),并加入超纯水 35mL,然后利用涡旋振荡 15min。把装有土壤悬浊液的离心管转移到摇床进行同平面回旋振荡(200r/min,1 200min)。振荡结束后,离心管中的土壤沉淀随之分为不同颜色的两层:上层为黑色的 SGO 薄层;下层为土灰色的土壤厚层,如图 8-1A 所示。将离心管里水相倒掉后,用不锈钢药勺把沉淀上层湿 SGO 转移到新的离心管中,合并 20 个离心管的 SGO。然后把收集的粗 SGO 用上述方法二次分离作为纯化操作。最后,用去离子水反复冲洗纯化后的湿 SGO 并用 0.1μm 水系滤膜进行抽滤收集 SGO。滤膜收集到的 SGO 进行冷冻干燥获得 SGO 粉末,如图 8-1B,然后分析和表征 SGO 粉末。为了抵消在样品分离过程中的样品变化,把 PGO 分散到水中制成悬浊液并重复上述的 SGO 分离过程,最后获得 PGO 粉末。

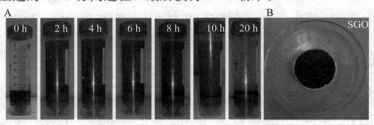

图 8-1 SGO 的分离过程
A. 振荡不同时间后的分离效果　B. 分离出的 SGO 粉末

8.2.2 土壤细菌菌群分析过程

主要步骤包括：DNA 提取、PCR 反应、高通量测序、序列加工处理和生物信息学分析。PCR 反应扩增细菌 16S rRNA V4~V5 区的 DNA 片段，PCR 扩增引物：515F（5′-barcode-GTGCCAGC-MGCCGCGG-3′）和 907R（5′-CCGTCAATTCMTTTRAGTTT-3′）。原始序列已经上传至 NCBI Sequence Read Archive 数据库（Accession Number：SRP050357）。

对于有效序列，使用 UPARSE（version 7.1）将相似度超过 97%以上的归类到一个 OTU，利用 UCHIME 去除 chimeric sequences。在聚类信息的基础上，通过 MOTHUR 获得每个样品的稀释曲线（rarefaction curves）、菌群丰富度指数 Chao、ACE 和菌群多样性指数 Shannon。使用 RDP Classifier 并对比 silva（SSU115）16S rRNA database 对 16S rRNA 基因序列进行系统发育分析，置信范围为 70%。最终将基因序列系统分类到门（phylum）、纲（class）、目（order）、科（family）和属（genus）5 个水平。利用 R Project 软件进行层序聚类分析。通过绘制文氏图描述样品间的相似性和差异性。

8.2.3 分析鉴定 SGO 表面固定的小分子有机物的方法

以甲醇：氯仿：水＝2.5：1：1（体积比）作为提取溶剂，提取方法采用微波辅助萃取，然后用衍生化试剂 MSTFA 和 BSTFA 衍生化后进行气相色谱串联质谱（GC-MS/MS）分析。具体步骤参考第 6 章中 LGO 表面吸附的根系分泌物的分离鉴定方法。

8.2.4 定性表征 PGO 和 LGO 的方法

本文对 PGO 和 LGO 的分析表征手段包括：扫描电子显微镜（SEM）、透色电子显微镜（TEM）、原子力显微镜（AFM）、zeta 电位、粒径分布、X-射线光电光谱仪（XPS）、傅里叶变换红外光谱仪（FTIR）、拉曼光谱分析（Raman）、紫外可见光谱分析

第8章 氧化石墨烯在土壤中性质改变和引起的土壤细菌菌群变化

(UV-vis)、电子顺磁共振（EPR）和氧化还原电位（Eh），具体的方法步骤参考第6章对LGO的表征。

8.2.5 数据处理和分析

除了在文中特别说明，所有实验处理都重复3次，误差棒表示标准差。标准差通过SPSS 19计算而得。

本章中，"PGO"和"SGO"分别代表原始氧化石墨烯和经土壤修饰后的氧化石墨烯。

8.3 土壤细菌菌群的丰富度和多样性

通过高通量测序，为伴有PGO的土壤样品PGOS和土壤样品CS的微生物菌群建立了两个16S rRNA基因文库，并分别获得了17 882条和15 651条高质量序列。16S rRNA序列的相似性高于97%定义为一个OTU，PGOS和CS分别生成702个和691个OTU。通过对两个样品进行多样性分析，结果见表8-1。Chao指数用来估算样品中OTU的总数，经过计算，PGOS（731）高于CS（713）。OTU和Chao的结果说明，在土壤中加入PGO后，微生物菌群的丰富度（richness）略有上升，稀释曲线（rarefaction curves，图8-2）的结果也证实了这个结果。Shannon指数是衡量细菌菌群多样性的指标，可同时反映其菌群物种丰富度和均匀度[229]。Shannon指数的结果也显示，PGOS（5.15）略高于CS（5.12）。文氏图用来描述两组OTU的相同点和不同点，如图8-3A所示。两个样品共有764个OTU，629个OTU是相同的，73个OTU只属于PGOS，62个OTU只属于CS，文氏图的结果也可以看出PGOS较CS菌群菌种更丰富。经过分类鉴定，如图8-3B和图8-3C所示，两组样品虽各独有少量OTU，但都可归入相同的菌门，例如Proteobacteria、Chloroflexi、Bacteroidetes、Planctomycetes、Gemmatimonadetes、Acidobacteria、Actinobacteria和Verrucomicrobia。PGOS和CS共有的菌门分别占到89.04%和82.26%。

表 8-1 PGOS 和 CS 多样性指数统计

项目	Reads	3% distance			
		OTU	ACE	Chao	Shannon
PGOS	17 882±231	702±5	728±4	731±9	5.15±0.01
CS	15 651±173	691±3	715±5	713±6	5.12±0.01

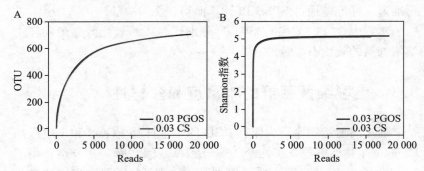

图 8-2 Rarefaction 曲线和 Shannon-Wiener 曲线

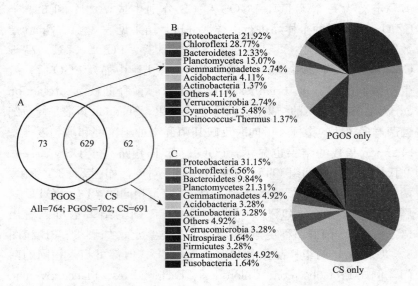

图 8-3 PGOS 和 CS 两组 OTU 的相同点和不同点
A. 文氏图 B. PGOS 独有 OTU 的门水平归属 C. CS 独有 OTU 的门水平归属

第8章 氧化石墨烯在土壤中性质改变和引起的土壤细菌菌群变化

8.4 土壤细菌菌群的分类组成

为了细致地分析 PGO 进入土壤后所带来的细菌菌群变化，本研究将样品 OTU 鉴定到门、纲和属 3 个分类水平，如图 8-4 所示，PGOS 和 CS 两组菌群的优势菌门、菌纲和菌属一致。在门水平，在主要的 10 个菌门中，两个样品相同的占 8 个（例如，Proteobacteria、Chloroflexi、Bacteroidetes 和 Planctomycetes 等），分别占 PGOS 和 CS 所有 OTU 的 89.40% 和 95.41%。在纲水平，在 18 个优势菌纲中，两个样品相同的占 15 个（例如，Acidobacteria、Actinobacteria、Alphaproteobacteria、Anaerolineae 和 Bacilli 等），分别占 PGOS 和 CS 所有克隆数的 83.34% 和 89.85%。在属水平，在 19 个优势菌属中，两个样品相同的占 8 个（例如，*Arthrobacter*、*Devosia* 和 *Lactococcus* 等），分别占 PGOS

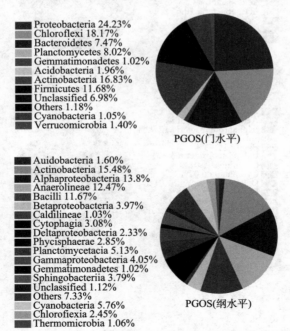

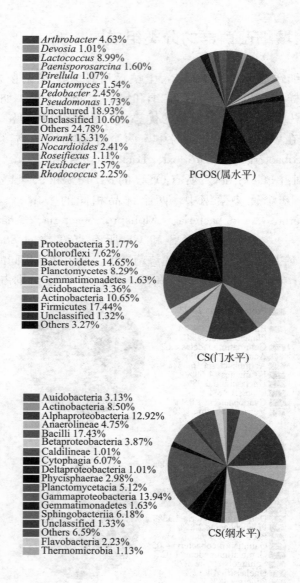

第8章 氧化石墨烯在土壤中性质改变和引起的土壤细菌菌群变化

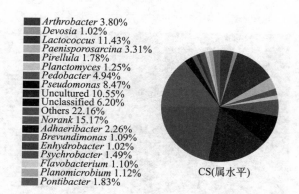

图 8-4 PGOS 和 CS 细菌菌群在门、纲和属水平的组成分析
注：相对丰度低于 1% 归为 "Others"。

和 CS 所有克隆数的 23.03% 和 36.00%。以上结果说明 PGOS 和 CS 菌群在菌门和菌纲两个分类级别差异不大，而在菌属分类上差异明显。

只在 PGOS 出现而在 CS 没有出现的菌属中，*Nocardioides*（2.41%）和 *Rhodococcus*（2.25%）是革兰氏阳性菌，*Roseiflexus*（1.11%）和 *Flexibacter*（1.57%）是革兰氏阴性菌。在 PGOS 没有出现而只在 CS 出现的菌属中，*Planomicrobium*（1.12%）属于革兰氏阳性菌，*Adhaeribacter*（2.26%）、*Brevundimonas*（1.09%）、*Enhydrobacter*（1.02%）、*Psychrobacter*（1.49%）、*Flavobacterium*（1.10%）和 *Pontibacter*（1.83%）属于革兰氏阴性菌。这些发现表明，在 PGO 进入土壤后所引起的细菌菌群改变依赖于菌种的差异而不受革兰氏分类差异的影响。

为了进一步了解菌群结构和功能的变化，本研究在菌属分类水平上对丰度前 100 的菌属进行了层序聚类分析，如图 8-5 所示。那些在 PGOS 比 CS 丰度高的菌属分布于很多不同的菌门。在这些菌属中，革兰氏阳性菌（*Rhodococcus*、*Nocardioides* 和

· 121 ·

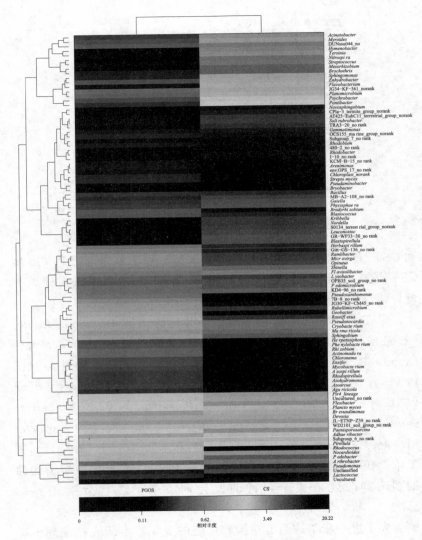

图 8-5 菌群层序聚类分析结果

注：相对丰度低于 1% 归为"Others"。

Actinomadura)都属于同一菌门 Actinobacteria,革兰氏阴性菌却属于不同的菌门(Proteobacteria、Chloroflexi、Verrucomicrobia、Bacteroidetes 和 Planctomycetes)。在 PGOS 比 CS 中丰度低的菌属中,革兰氏阳性菌属于 Actinobacteria(*Kribbella* 和 *Solirubrobacter*)和 Firmicutes(*Brochothrix*、*Streptococcus* 和 *Planomicrobium*),革兰氏阴性菌属于 Proteobacteria 和 Planctomycetes。通过以上结果可以总结出,与 CS 相比,在 PGOS 中丰度下降的革兰氏阳性菌属于 Actinobacteria 和 Firmicutes,而丰度升高的革兰氏阳性菌却只属于 Actinobacteria。所以,在 PGO 进入土壤后,细菌菌群丰度的变化也只依赖于细菌菌种的特性而不受革兰氏分类特性的差异所影响。有研究认为,在水环境中,石墨烯类纳米材料的抑菌活性依赖于细菌菌种而不依赖于细菌革兰氏的分类特性[312]。本研究的结果证实在土壤环境中有相似的结果。

此外,具有固氮作用的属于变形菌门(Proteobacteria)的菌群(包括 *Azoarcus*、*Azospirillum*、*Bradyrhizobium*、*Herbaspirillum*、*Ensifer*、*Microvirga* 和 *Rhizobium*)在 PGOS 中呈现出比 CS 更高的丰度。其中,有些固氮菌,例如 *Azospirillum* 和 *Azoarcus*,在缺氧状态下还能够异化硝酸盐为亚硝酸盐或者 N_2O 和 N_2。*Geobacter* 也是一类具有异化代谢能力的细菌,在 PGOS 比在 CS 呈现出更高的丰度。*Geobacter* 是异化铁还原菌,与上述讨论的固氮菌同属于变形菌门。此外,一组多环芳烃降解菌群在 PGOS 中表现出比 CS 更高的丰度,例如 *Sphingobium*、*Rhodococcus* 和 *Pseudoxanthomonas*[393]。综上所述,在 PGO 进入土壤后,细菌菌群尤其在菌属水平上有选择性地变得丰富。

8.5 氧化石墨烯表面形貌变化

上文讨论的 PGO 进入土壤后所引起的细菌菌群变化与 PGO 被土壤修饰而性质发生改变是相伴发生的。通过 SEM、TEM 和

AFM 表征手段研究 PGO 被土壤修饰前后的形貌变化。SEM 表征结果见图 8-6A 和图 8-6B，PGO 和土壤修饰后的氧化石墨烯（SGO）都呈现纳米片层形貌，但是 SGO 比 PGO 表面粗糙，一些颗粒状物质分布于 SGO 片层表面或被片层包埋。如 TEM（图 8-6C 和图 8-6D）结果所示，SGO 的透明度明显低于 PGO，而且在 SGO 表面和内部分散着尺寸不均一的黑色斑块。通过 EDS 能谱分析，如图 8-6E 所示，这些黑斑的化学成分中除了氧化石墨烯所必须含有的 C 和 O 以外，还包括 N、Mg、Al、Si、K、Ca 和 Fe，这些元素来源于土壤。AFM 用来表征纳米材料的厚度，如图 8-7 所示，PGO 呈现的厚度 0.862 nm 符合单层氧化石墨烯的要求[347]。SGO 厚度不均，厚的区域为 7.162nm，薄的区域为 3.562nm。对于以上形貌的表征分析发现，SGO 纳米片的 AFM 图中厚的区域与 TEM 图中的黑斑、SEM 图中的颗粒相吻合。

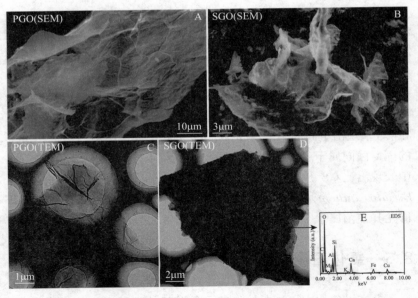

图 8-6　纳米材料的形貌表征：SEM 和 TEM（EDS）表征结果

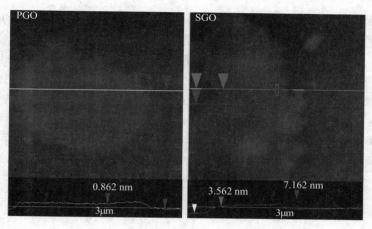

图 8-7 纳米材料 AFM 表征结果

8.6 氧化石墨烯表面化学官能团的变化

本研究通过 XPS 分析 PGO 和 SGO 表面化学性质，全谱分析见图 8-8。PGO 经全谱分析，包含 69.73% C1s 和 30.27% O1s。SGO 经分析包含 41.35% C1s、44.15% O1s、1.32% N1s 和 13.18% 的其他元素（例如，Mg、Al、Si、K、Ca 和 Fe），这个表征结果与 EDS（图 8-6E）能谱分析结果一致。这些无机元素来源于土壤矿质成分，有研究报道 PGO 可以通过静电吸附作用吸附矿物质[385]。图 8-9 呈现了 C、O 和 N 三种元素的精细谱分析结果。PGO 的 C1s 谱图组成包括未氧化的 44.98% C—C/C=C、28.56% C—O（代表了羟基和环氧基官能团）和 26.46% C=O。PGO 的 C1s 谱图组成包括 C—C/C=C（69.09%）、C—O（17.96%）、C=O（7.60%）和 C—N（5.36%），与 FTIR 的分析结果一致，见图 8-10。经过在土壤中的交互作用，PGO 表面 C—O 和 C=O 的数量明显下降，分别从 28.56% 和 26.46% 下降到 17.96% 和 7.60%。有研究表明，PGO 通过与环境微生物 *Escherichia coli*[327] 和 *Shewanella*[321,328,394] 交互作用，含氧官能团减少。

细菌转化 PGO 的潜在机制与 *Shewanella* 呼吸代谢和 *Escherichia coli* 糖酵解代谢有关，而且在还原反应过程中，PGO 作为最终的电子受体参与其中[312]。此外，也有报道其他微生物，例如酵母菌和植物内生菌，可以还原 PGO[326,391]。土壤是微生物的资源库和大本营，土壤中分布的微生物数量巨大，种类繁多，具有多种代谢途径。PGO 在土壤中被土壤微生物还原是不可避免的。另外，通过 FTIR（图 8-10）分析，在 SGO 表面检测到亚甲基（Methylene），说明通过在土壤中的修饰作用，在 SGO 表面引入了碳链，提高了 C/O。

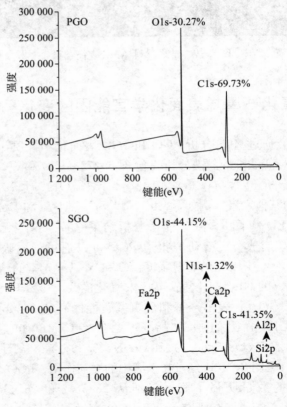

图 8-8　纳米材料 XPS 全谱表征结果

第 8 章 氧化石墨烯在土壤中性质改变和引起的土壤细菌菌群变化

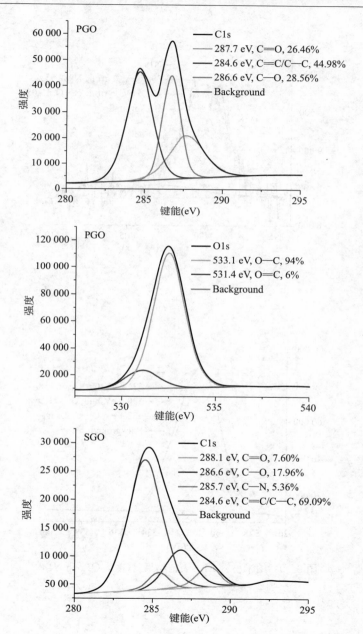

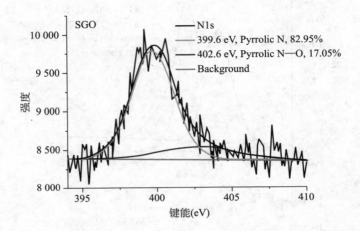

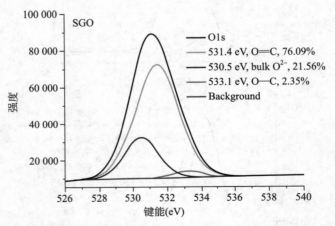

图 8-9 纳米材料 XPS 表征结果（C1s、O1s 和 N1s）

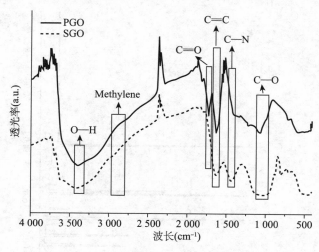

图 8-10 纳米材料的 FTIR 表征结果

与 PGO 相比，N1s 信号的检出说明在 SGO 表面掺入新的 N 元素。经过 N 元素的精细谱分析，N1s 由 pyrrolic N（82.95%）和 pyridinic N—O（17.05%）组成，两种官能团都没有固定在石墨烯结构平面内，只是固定在平面的边缘、表面或者缺陷位置[352,395]。N1s 谱图中的 pyridinic N—O 可能来源于土壤中广泛存在的硝酸盐或者亚硝酸盐，其检出也与 SGO 的 O1s 谱图中 bulk O^{2-} 的检出相吻合。SGO 表面的 pyrrolic N 可能来源于土壤中的含氮有机物，也可能来源于土壤微生物的反硝化代谢过程[330]。

8.7 分析鉴定氧化石墨烯表面固定的小分子有机物

本研究通过 GC-MS/MS 分析鉴定固定在 SGO 表面的小分子有机物，共检测出 14 种小分子有机物，如图 8-11 所示。这些有机小分子包括有机酸类、酚类、苯类、醚类、酰胺类、烷烃类和醛酮类，它们可能来自土壤中的有机物或者土壤微生物的代谢。在检测出的 14 种有机小分子中，其中 8 种能在所采集的干土中检测到（表 8-2），这 8 种物质分别是苯酚（Phenol）、二十八烷（Octaco-

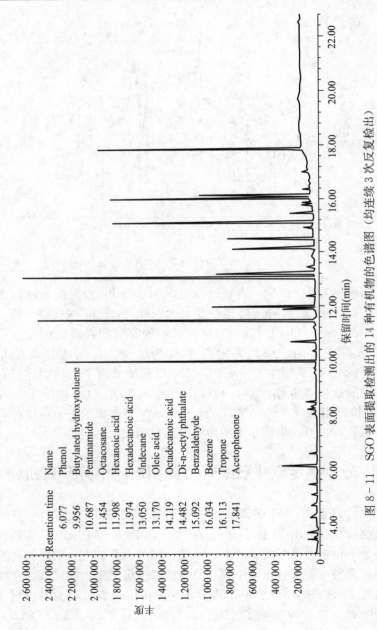

图 8-11 SGO 表面提取检测出的 14 种有机物的色谱图（均连续 3 次反复检出）

第8章 氧化石墨烯在土壤中性质改变和引起的土壤细菌菌群变化

sane)、己酸（Hexanoic acid）、棕榈酸（Hexadecanoic acid）、十一烷（Undecane）、硬脂酸（Octadecanoic acid）、苯（Benzene）和苯甲醛（Benzaldehyde）。PGO 含有多种含氧官能团（—OH、—COOH和—CHO）和多种特殊的纳米结构，如纳米孔、活性边缘、悬空碳和 sp^2 结构[342,355]。上文提到的有机小分子含有羧基、酚羟基或者苯环，这些官能团可以通过氢键和 π-π 键与 PGO 联结。这些在 SGO 表面固定的有机小分子同时提高了 C/O。而且，戊酰胺（Pentanamide）的检出也与 XPS 表征的 N1s 谱图结果相吻合。

表 8-2 原土中分析检测出的有机小分子物质

序号	物质名称	序号	物质名称
1	Benzene	19	Tetradecanoic acid
2	Butanoic acid	20	d-Galactose
3	Benzoic acid	21	Hexanoic acid
4	Propanoic acid	22	D-Mannitol
5	3-Methylbutyl	23	Tetratetracontane
6	Glycine	24	Heptacosane
7	Amino levulinic acid	25	2-Propenoic acid
8	Glycerol	26	Hexadecanoic acid
9	Hexadecane	27	Eicosane
10	Octacosane	28	Hentriacontane
11	Octane	29	Docosane
12	L-(+)-Lactic acid	30	Heptadecanoic acid
13	Benzaldehyde	31	Octadecanoic acid
14	1-Butanol	32	Benzeneethanamine
15	Phenol	33	Decane
16	Undecane	34	2-Pentadecanone
17	Phosphonic acid	35	Dotriacontane
18	Pentacosane	36	Octadecane

(续)

序号	物质名称	序号	物质名称
37	Propenamide	50	Propionic acid
38	N-Methyl-1-adamantaneacetamide	51	D-glucopyranoside
39	Ethanedioic acid	52	Phosphenimidous amide
40	2-Phenylbutyric acid	53	Hexadecanamide
41	2-Butenoic acid	54	Anthracene
42	d-Mannose	55	Octadecanamide
43	Tetracosane	56	Nonadecane
44	2-Cyclopenten-1-one	57	5-Pentadecanone
45	Monopalmitin	58	1-Octacosanol
46	Myristic acid	59	Cyclohexane
47	Aniline	60	L-Proline
48	Benzamide	61	1-Penten-3-one
49	Butane		

注：61种有机物均连续3次重复检出。

8.8　氧化石墨烯粒径分布和表面电荷的变化

纳米材料的粒径分布和表面电荷对于其环境行为和相关的生态风险起主要作用[147,349,396]。如图8-12所示，PGO粒径分布于342～532nm之间；SGO粒径分布于142～396nm之间，说明在土壤中经过老化，PGO尺寸减小。有研究发现环境微生物和辣根过氧化物酶（Horseradish peroxidase，HRP）可以生物降解PGO[312,397]。对PGO而言，在土壤环境中相似的生物降解过程也可能发生，从而改变PGO尺寸，因为土壤微生物种类繁多以及其分泌各种各样的酶。不同pH条件的zeta测试结果见图8-13，在所测pH范围（pH 5～9），PGO和SGO都带有负电荷。PGO所携带负电荷多于SGO，说明PGO比SGO更稳定。

第8章 氧化石墨烯在土壤中性质改变和引起的土壤细菌菌群变化

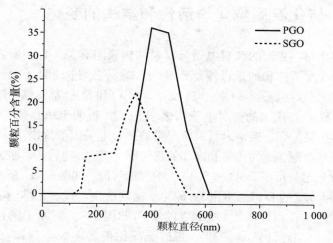

图 8-12 纳米材料的粒径分布分析结果

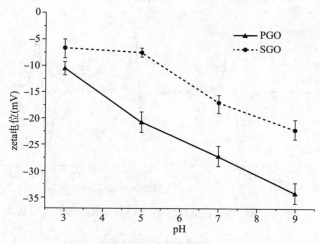

图 8-13 纳米材料的 zeta 电位分析结果

8.9 氧化石墨烯化学活性和结构的变化

纳米材料表面的孤对电子与材料结构缺陷有关，直接影响纳米材料的化学活性和纳米毒性[349,355,356]。本研究通过 EPR 手段分析纳米材料的孤对电子，如图 8-14 所示，PGO 和 SGO 都能够产生孤对电子。SGO 呈现出的孤对电子稍多于 PGO，说明土壤修饰 PGO 后的化学活性增强。碳纳米材料所产生的孤对电子因不规则结构所致，例如，悬空键、碳面边缘和内部的纽结等[147,358]。随后，本研究通过 Raman 光谱分析了 PGO 和 SGO 的不规则结构，如图 8-15 所示。位于 1 363cm^{-1} 的 D 带通常由 sp^3 碳原子振动产生，代表了结构缺陷和不规则的结构；位于 1 593cm^{-1} 的 G 带归因于 sp^2 碳原子的伸长振动[325,398]。通过计算 D 带和 G 带的强度比值（D/G），SGO（1.44）高于 PGO（1.15），说明经过土壤修饰，PGO 结构的不规则程度增强，这个结果与孤对电子的分析结果相吻合。

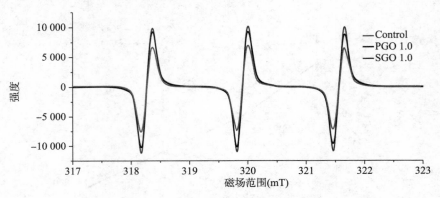

图 8-14 纳米材料的 EPR 表征结果

本研究通过 UV-vis 光谱分析来反映纳米材料的光化学活性，如图 8-16 所示。PGO 和 SGO 的最大吸收峰分别位于 203nm 和 196nm，分别反映了 n-π 和 π-π 跃迁（E2 带）[360]。SGO 谱图 200nm 附近的最大吸收峰远高于 PGO 谱图，而此峰由土壤中固定

第8章 氧化石墨烯在土壤中性质改变和引起的土壤细菌菌群变化

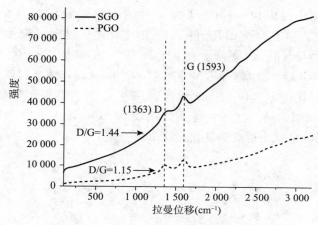

图 8-15 纳米材料的 Raman 表征结果

的有机小分子所携带的—COOH、—NH—、不饱和酮基和孤对电子所产生。与 PGO 相比，SGO 的最大吸收峰从 203nm 转移到 196nm，说明 SGO 电子浓度降低，这是由于 SGO 结构不规则程度提高后 sp^3 碳原子增多所致[399]。另外，与 PGO 相比，SGO 在 205～800nm 区域吸收强度明显降低，说明在固定了土壤中的化学成分后其石墨烯片层内的电子共轭作用被打乱[361]。

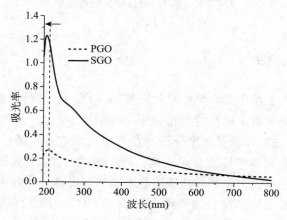

图 8-16 纳米材料的 UV-vis 表征结果

氧化还原电位（Eh）作为一种纳米材料化学活性的指标，PGO 和 SGO 在不同 pH 条件下 Eh 的变化，见图 8-17。随 pH 升高，两种材料的 Eh 下降。在 pH 3～11，SGO 的 Eh 值始终高于 PGO。PGO 经土壤修饰后尺寸减小（图 8-12），纳米材料尺寸越小，其氧化还原电位和化学活性越高[147]。另外，经土壤修饰后，SGO 表面引入新的化学官能团，例如 N—O（图 8-9，N1s 谱），致使 SGO 呈现出更高的氧化还原电位。

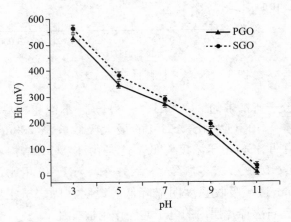

图 8-17 纳米材料的 Eh 分析结果

8.10 小结

本章向土壤中加入原始氧化石墨烯（PGO）进行土培实验后，研究了细菌菌群多样性的变化和 PGO 的性质变化。首先提取细菌 16SrRNA 进行高通量测序分析，发现加入氧化石墨烯后土壤细菌菌群丰富度和多样性较对照提高。PGO 的加入改变了土壤细菌菌群结构，一些固氮菌和异化铁还原菌丰富度提高。然后从土壤处理组分离出经土壤修饰后的氧化石墨烯（SGO），并进行表征，分析其性质变化。SGO 较 PGO 厚度增加，元素 C/O 升高，质地更粗

第 8 章　氧化石墨烯在土壤中性质改变和引起的土壤细菌菌群变化

糙，透明度下降，尺寸变小。通过 XPS 和 EDS 表征分析，SGO 表面引入新的含氮官能团和新的无机元素 Mg、Al、Si、K、Ca 和 Fe 等。SGO 表面官能团的分析与表面吸附的有机小分子一致。SGO 比 PGO 表面携带更少的负电荷，比 PGO 更不稳定。另外，SGO 展现出更高的化学活性，例如，孤对电子更多，不规则结构程度更高。

第 9 章 石墨烯基纳米材料对土壤细菌菌群的影响研究

9.1 引言

土壤生态系统是各种纳米材料的最终归宿。如今,石墨烯基纳米材料(Graphene-based nanomaterials,GBNs)在农业领域的应用和产品呈现快速增长趋势,其在土壤环境的积累不可避免[400,401]。有研究认为,GBNs 在农业领域的应用浓度已经达到了很高水平[402,403]。GBNs 进入土壤将有可能扰乱土壤微生物菌群结构,进而影响地球生态系统中微生物代谢参与调控的生物地球化学元素循环[166]。前人的研究主要关注 GBNs 对微生物生物量、多样性和种群的影响。然而,前人的研究结果仍然不一致[404]。有时,很小的实验条件差异也有可能导致检测到的微生物种群信息的变化。反而,微生物菌群功能组比种群信息更稳定[405]。本研究比较研究了了氧化石墨烯(Graphene oxide,GO)和还原氧化石墨烯(Reduced graphene oxide,RGO)对土壤细菌菌群多样性和功能组的影响。

9.2 土壤培育实验方案与设计

本研究所用 GO 和 RGO,均购自南京先丰纳米材料有限公司,其理化特性见表 9-1。

表 9-1 GBNs 理化特性

种类	纯度(%)	片径(μm)	C(%)	O(%)	层级
GO	>99	0.5~5	71.23	28.77	单层
RGO	>99	0.5~5	95.97	4.03	单层

第9章 石墨烯基纳米材料对土壤细菌菌群的影响研究

本研究采用的土壤样品属于棕壤（36°04′N、111°34′E），其理化性状见表9-2。

表9-2 土壤理化性状

pH	总N（%）	总P（%）	总K（%）
8.14	1.54	1.11	6.21

9.2.1 土壤培育实验设计

将GBNs按1.0g/L的浓度分散到去离子水中，通过超声处理后使其形成稳定的悬浊液，具体的步骤详见参考文献[392]。实验用土来自山西省临汾市尧都区周边农田的表层土（0～20cm）。土壤风干后，用1mm孔径的尼龙网筛去除石头、土块等杂质。称取去除杂质的干土1 500g转移到塑料花盆（平均直径18cm、高12cm）中，浇水并调整到田间持水量的50%。然后将花盆放在25℃条件下预培养7d，恢复土壤微生物活性。培养结束后，将提前准备好的GO悬浊液与花盆中的土样充分搅拌混合，使GBNs最终浓度为50mg/kg（干土）。尽管GBNs在土壤环境中的具体浓度还不确定，但是，本研究基于已经发表的前人研究和目前的工业发展选择了最终的暴露浓度。到2017年底，GBNs在农业领域的应用浓度已经达到很高水平[402]。对照组花盆加入相同体积的去离子水。最后把所有花盆放到温室中培养（25℃）。其间，用去离子维持土壤水分保持在田间最大持水量的50%。土培实验在温室进行90d后，从花盆的底部、中部和上部分别取样并混合，收集的土样用于土壤微生物菌群分析。本研究共设置3个处理组，每个处理组包括3个重复。Control组：对照组，没有添加GO或RGO；GO组：添加GO的处理；RGO组：添加RGO的处理。

9.2.2 土壤细菌菌群分析

主要步骤包括：DNA提取、PCR扩增、高通量测序、序列加工处理和生物信息学分析。通过PCR反应扩增细菌16S rRNA V3～V4区的DNA片段，PCR扩增引物：338F（5′-ACTC-

CTACGGGAGGCAGCA-3′）和 806R（5′-GGACTACHVGGGT-WTCTAAT-3′）。扩增产物在 Illumina MiSeq 平台进行高通量测序。原始序列已经上传至 NCBI Sequence Read Archive（SRA）数据库（Accession Number：PRJNA552951）。最终将基因序列系统分类到门（phylum）、纲（class）、目（order）、科（family）和属（genus）5 个水平。通过 QIIME 获得每个样品的 Chao1 指数、ACE 指数和 Shannon 指数。利用 R Project 软件进行层序聚类分析。样品间的菌群结构变化通过 UniFrac 举例度量，并通过主坐标（PCoA）分析图呈现。菌群分类信息通过 FAPROTAX 进行功能注释预测。FAPROTAX 取词自 Functional Annotation of Prokaryotic Taxa，是 Louca 等人为解析微生物群落功能于 2016 年创建的基于原核微生物分类的功能注释数据库[166]。FAPROTAX 是基于目前对可培养菌的文献资料手动整理的原核功能注释数据库，较适于对环境样本（如海洋、湖泊等）的生物地球化学循环过程（特别是碳、氢、氮、磷、硫等元素循环）进行功能注释预测。

9.2.3 数据分析

所有实验都进行 3 次重复，所得的数据均进行方差分析，文中误差棒代表了标准差，结果以 SPSS 19.0 统计软件进行分析，数据的显著水平均指 $P<0.05$（ANOVA，LSD）。

9.3 石墨烯基纳米材料对土壤细菌菌群 α-多样性的影响

土培实验结束后，经 DNA 提取、PCR 扩增和高通量测序后发现，三组（9 个）样品共获得 391 821 个高质量序列。Control、GO 和 RGO 实验组获得的平均序列数分别为 44 421 个、47 867 个和 38 318 个。ACE 和 Chao1 指数反映微生物菌群的种群丰富度，Shannon 指数反映种群多样性。如表 9-3 所示，Control 组的 ACE 和 Shannon 指数最高，说明 GO 和 RGO 添加到土壤都引起了土壤细菌菌群丰富度和多样性的降低。Shannon 指数在 Control、GO 和 RGO 实验组之间的差异不显著（$P>0.05$）。RGO 实验组的 ACE 和 Chao1 指数显著（$P<0.05$）低于 Control 和 GO 实验组，说明 RGO 的暴露明显引起了土壤细菌菌群丰富度的降低。

表 9-3 不同处理的土壤细菌菌群丰富度和多样性

处理	Sequence	3% distance		
		ACE	Chao1	Shannon
Control	44 421±1 542ab	3 803±301a	3 461±283a	10.36±0.46a
GO	47 867±7 348a	3 777±709a	3 597±594a	9.58±1.46a
RGO	38 318±1 036b	2 349±107b	2 318±53b	10.32±0.37a

注：不同小写字母表示在 0.05 水平（LSD 检验）差异显著，$n=3$。

9.4 石墨烯基纳米材料对土壤细菌菌群组成的影响

基于 97% 的相似性，OTU 在 Greengenes 数据库比对分析结果如图 9-1 所示，Control、GO 和 RGO 组的优势菌群在门水平是一致的，包括：Proteobacteria、Actinobacteria、Chloroflexi、Acidobacteria、Gemmatimonadetes、Cyanobacteria、Firmicutes、Nitrospirae、Tectomicrobia、Bacteroidetes、Planctomycetes 和 Parcubacteria。GO 和 RGO 的暴露没有引起土壤细菌优势菌群在门水平的明显变化。

PCoA 分析用于描绘不同实验组的差异和相似度，图中每个点代表 1 个样本，不同颜色的点属于不同实验组，两点之间的距离越近，表明两个样本之间的细菌群落结构相似度越高，差异越小。如图 9-2 所示，图中的 9 个样本趋于 3 个集群，说明不同处理对细菌菌群结构产生了明显的影响（PERMANOVA，$P<0.05$）。PC1 轴把 GO 组和 Control 组菌群分开，差异性为 30.23%；PC2 轴把 RGO 组和 Control 组菌群分开，差异性为 18.56%。PCoA 图的结果说明，GO 所引起的土壤细菌菌群结构变化大于 RGO。

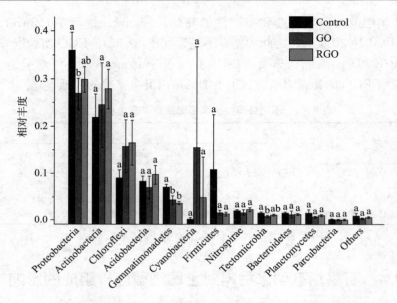

图 9-1　不同处理土壤优势菌群在门水平的相对丰度

注：误差棒代表平均值±标准差，不同小写字母表示在 0.05 水平（LSD 检验）差异显著，$n=3$。"Others"代表的菌门至少在 1 组处理的平均相对丰度<0.1%。

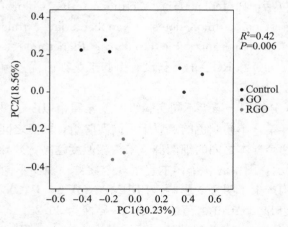

图 9-2　不同处理土壤细菌菌群的 PCoA 分析

注：实验组间在 0.05 水平差异显著（基于 PERMANOVA 分析，置换检验次数：999）。

9.5 石墨烯基纳米材料对土壤细菌菌群功能组的影响

FAPROTAX 数据库适用于对环境样本的生物地球化学循环过程（特别是碳、氢、氮、磷和硫等元素循环）进行功能注释预测[166]。因其基于已发表验证的可培养菌文献，其预测准确度可能较好。FAPROTAX 数据库中总共注释了 90 个功能组，本研究的样本数据共能匹配 74 个功能组，见表 9-4。

表 9-4 不同处理土壤细菌菌群功能组的相对丰度
（基于 FAPROTAX 数据库）

功能组	Control	GO	RGO
chemoheterotrophy	0.162±0.028a	0.117±0.005b	0.125±0.015b
aerobic_chemoheterotrophy	0.115±0.028a	0.114±0.005a	0.12±0.014a
fermentation	0.068±0.048a	0.01±0.002b	0.015±0.002ab
nitrification	0.026±0.003a	0.025±0.003a	0.026±0.003a
aerobic_nitrite_oxidation	0.014±0.002a	0.014±0.002a	0.015±0.001a
aerobic_ammonia_oxidation	0.012±0.002a	0.011±0.001a	0.011±0.003a
nitrate_reduction	0.012±7×10^{-4}a	0.01±1×10^{-3}b	0.01±9×10^{-4}b
animal_parasites_or_symbionts	0.012±0.005a	0.008±7×10^{-4}a	0.009±0.003a
respiration_of_sulfur_compounds	0.01±3×10^{-4}a	0.006±5×10^{-4}b	0.007±4×10^{-4}c
human_pathogens_all	0.008±0.001a	0.008±7×10^{-4}a	0.009±0.003a
ureolysis	0.007±0.001a	0.004±7×10^{-4}b	0.006±0.002ab

(续)

功能组	Control	GO	RGO
human_pathogens_pneumonia	$0.006\pm0.001a$	$0.007\pm5\times10^{-4}a$	$0.008\pm0.002a$
sulfur_respiration	$0.007\pm9\times10^{-4}a$	$0.006\pm8\times10^{-4}b$	$0.006\pm6\times10^{-4}ab$
nitrogen_fixation	$0.006\pm0.002a$	$0.003\pm4\times10^{-4}b$	$0.005\pm0.001ab$
predatory_or_exoparasitic	$0.006\pm0.001a$	$0.005\pm4\times10^{-4}a$	$0.004\pm0.001a$
manganese_oxidation	$0.004\pm0.001a$	$0.005\pm8\times10^{-4}a$	$0.004\pm7\times10^{-4}a$
phototrophy	$0.004\pm0.001a$	$0.023\pm0.017a$	$0.024\pm0.036a$
nitrogen_respiration	$0.003\pm7\times10^{-4}a$	$0.002\pm2\times10^{-4}b$	$0.003\pm6\times10^{-4}ab$
nitrate_respiration	$0.003\pm7\times10^{-4}a$	$0.002\pm2\times10^{-4}b$	$0.003\pm6\times10^{-4}ab$
photoautotrophy	$0.003\pm9\times10^{-4}a$	$0.022\pm0.017a$	$0.023\pm0.036a$
cellulolysis	$0.002\pm6\times10^{-4}a$	$0.003\pm3\times10^{-4}a$	$0.002\pm0.001a$
photoheterotrophy	$0.002\pm5\times10^{-4}a$	$0.002\pm3\times10^{-4}a$	$0.002\pm5\times10^{-4}a$
aromatic_compound_degradation	$0.003\pm9\times10^{-4}a$	$0.008\pm0.001b$	$0.006\pm0.002ab$
sulfate_respiration	$0.003\pm6\times10^{-4}a$	$5\times10^{-4}\pm3\times10^{-4}b$	$0.001\pm1\times10^{-3}b$
nitrite_respiration	$0.002\pm4\times10^{-4}a$	$0.001\pm2\times10^{-4}a$	$0.001\pm5\times10^{-4}a$
human_gut	$0.004\pm0.005a$	$2\times10^{-5}\pm3\times10^{-5}a$	$8\times10^{-5}\pm4\times10^{-5}a$
mammal_gut	$0.004\pm0.005a$	$2\times10^{-5}\pm3\times10^{-5}a$	$8\times10^{-5}\pm4\times10^{-5}a$
nitrate_denitrification	$0.001\pm5\times10^{-4}a$	$0.001\pm2\times10^{-4}a$	$0.001\pm4\times10^{-4}a$

（续）

功能组	Control	GO	RGO
nitrite _ denitrification	$0.001 \pm 5 \times 10^{-4}$a	$0.001 \pm 2 \times 10^{-4}$a	$0.001 \pm 4 \times 10^{-4}$a
nitrous _ oxide _ denitrification	$0.001 \pm 5 \times 10^{-4}$a	$0.001 \pm 2 \times 10^{-4}$a	$0.001 \pm 4 \times 10^{-4}$a
denitrification	$0.001 \pm 4 \times 10^{-4}$a	$0.001 \pm 2 \times 10^{-4}$a	$0.001 \pm 4 \times 10^{-4}$a
anoxygenic _ photoautotrophy	$0.001 \pm 3 \times 10^{-4}$a	$0.001 \pm 2 \times 10^{-4}$a	$0.001 \pm 5 \times 10^{-4}$a
methylotrophy	$0.001 \pm 5 \times 10^{-4}$a	$9 \times 10^{-4} \pm 8 \times 10^{-5}$a	$0.001 \pm 4 \times 10^{-4}$a
anoxygenic _ photoautotrophy _ S _ oxidizing	$0.001 \pm 3 \times 10^{-4}$a	$0.001 \pm 2 \times 10^{-4}$a	$0.001 \pm 5 \times 10^{-4}$a
iron _ respiration	$0.001 \pm 5 \times 10^{-4}$a	$0.001 \pm 2 \times 10^{-4}$a	$0.001 \pm 1 \times 10^{-4}$a
hydrocarbon _ degradation	$8 \times 10^{-4} \pm 5 \times 10^{-4}$a	$7 \times 10^{-4} \pm 1 \times 10^{-4}$a	$0.001 \pm 5 \times 10^{-4}$a
cyanobacteria	$0.001 \pm 7 \times 10^{-4}$a	0.021 ± 0.018a	0.022 ± 0.036a
oxygenic _ photoautotrophy	$0.001 \pm 7 \times 10^{-4}$a	0.021 ± 0.018a	0.022 ± 0.036a
methanotrophy	$7 \times 10^{-4} \pm 5 \times 10^{-4}$a	$5 \times 10^{-4} \pm 5 \times 10^{-5}$a	$7 \times 10^{-4} \pm 3 \times 10^{-4}$a
chitinolysis	$8 \times 10^{-4} \pm 2 \times 10^{-4}$	$0.001 \pm 2 \times 10^{-4}$a	$8 \times 10^{-4} \pm 3 \times 10^{-4}$a
chloroplasts	$5 \times 10^{-4} \pm 2 \times 10^{-4}$	$0.001 \pm 6 \times 10^{-4}$a	0.001 ± 0.001a
intracellular _ parasites	$6 \times 10^{-4} \pm 9 \times 10^{-5}$a	$5 \times 10^{-4} \pm 6 \times 10^{-5}$a	$6 \times 10^{-4} \pm 2 \times 10^{-4}$a
dark _ oxidation _ of _ sulfur _ compounds	$4 \times 10^{-4} \pm 1 \times 10^{-4}$a	$3 \times 10^{-4} \pm 2 \times 10^{-5}$b	$4 \times 10^{-4} \pm 1 \times 10^{-4}$ab
dark _ hydrogen _ oxidation	$5 \times 10^{-4} \pm 5 \times 10^{-4} \pm$a	$9 \times 10^{-5} \pm 9 \times 10^{-5} \pm$a	$3 \times 10^{-4} \pm 2 \times 10^{-4} \pm$a
nitrite _ ammonification	$5 \times 10^{-4} \pm 4 \times 10^{-4}$a	$4 \times 10^{-5} \pm 4 \times 10^{-5}$a	$3 \times 10^{-5} \pm 4 \times 10^{-5}$a

(续)

功能组	Control	GO	RGO
dark_thiosulfate_oxidation	$3\times10^{-4}\pm5\times10^{-5}$a	$2\times10^{-4}\pm2\times10^{-5}$a	$3\times10^{-4}\pm1\times10^{-4}$a
fumarate_respiration	$5\times10^{-4}\pm5\times10^{-4}$a	$3\times10^{-5}\pm3\times10^{-5}$a	0 ± 0a
xylanolysis	$2\times10^{-4}\pm1\times10^{-4}$a	$3\times10^{-4}\pm1\times10^{-5}$a	$4\times10^{-4}\pm7\times10^{-5}$a
methanol_oxidation	$3\times10^{-4}\pm7\times10^{-5}$a	$4\times10^{-4}\pm1\times10^{-4}$a	$5\times10^{-4}\pm1\times10^{-4}$a
nitrate_ammonification	$1\times10^{-4}\pm9\times10^{-5}$a	$4\times10^{-5}\pm4\times10^{-5}$a	$3\times10^{-5}\pm4\times10^{-5}$a
human_pathogens_gastroenteritis	$4\times10^{-4}\pm5\times10^{-4}$a	0 ± 0a	0 ± 0a
human_pathogens_diarrhea	$4\times10^{-4}\pm5\times10^{-4}$a	0 ± 0a	0 ± 0a
knallgas_bacteria	$8\times10^{-5}\pm6\times10^{-5}$a	$3\times10^{-5}\pm3\times10^{-5}$ab	0 ± 0b
aromatic_hydrocarbon_degradation	$1\times10^{-4}\pm4\times10^{-5}$a	$2\times10^{-4}\pm2\times10^{-4}$a	$4\times10^{-4}\pm2\times10^{-4}$a
aliphatic_non_methane_hydrocarbon_degradation	$1\times10^{-4}\pm3\times10^{-5}$a	$2\times10^{-4}\pm1\times10^{-4}$a	$3\times10^{-4}\pm2\times10^{-4}$a
ligninolysis	$9\times10^{-5}\pm7\times10^{-5}$a	$6\times10^{-5}\pm2\times10^{-5}$a	$8\times10^{-5}\pm4\times10^{-5}$a
plant_pathogen	$5\times10^{-5}\pm2\times10^{-5}$a	$6\times10^{-5}\pm2\times10^{-5}$a	$2\times10^{-4}\pm2\times10^{-4}$a
aerobic_anoxygenic_phototrophy	$3\times10^{-5}\pm3\times10^{-5}$a	0 ± 0a	$1\times10^{-5}\pm2\times10^{-5}$a
dark_sulfide_oxidation	$4\times10^{-5}\pm4\times10^{-5}$a	0 ± 0a	$1\times10^{-5}\pm2\times10^{-5}$a
dark_iron_oxidation	$2\times10^{-5}\pm3\times10^{-5}$a	$2\times10^{-5}\pm3\times10^{-5}$a	$3\times10^{-5}\pm2\times10^{-5}$a

第9章 石墨烯基纳米材料对土壤细菌菌群的影响研究

（续）

功能组	Control	GO	RGO
invertebrate _ parasites	$5\times10^{-5}\pm2\times10^{-5}$a	$9\times10^{-5}\pm1\times10^{-5}$a	$1\times10^{-4}\pm5\times10^{-5}$a
thiosulfate _ respiration	$2\times10^{-5}\pm1\times10^{-5}$a	$3\times10^{-5}\pm1\times10^{-6}$a	$4\times10^{-5}\pm4\times10^{-5}$a
anammox	$8\times10^{-6}\pm1\times10^{-5}$a	$9\times10^{-6}\pm2\times10^{-5}$a	$3\times10^{-5}\pm4\times10^{-5}$a
human _ pathogens _ meningitis	$8\times10^{-6}\pm1\times10^{-5}$a	0 ± 0a	0 ± 0a
plastic _ degradation	$2\times10^{-5}\pm1\times10^{-5}$a	$9\times10^{-6}\pm2\times10^{-5}$a	$5\times10^{-5}\pm6\times10^{-5}$a
reductive _ acetogenesis	$2\times10^{-5}\pm1\times10^{-5}$a	$9\times10^{-6}\pm2\times10^{-5}$a	0 ± 0a
dark _ sulfite _ oxidation	$3\times10^{-5}\pm6\times10^{-5}$a	0 ± 0a	0 ± 0a
dark _ sulfur _ oxidation	$3\times10^{-5}\pm6\times10^{-5}$a	0 ± 0a	0 ± 0a
human _ pathogens _ septicemia	$2\times10^{-5}\pm3\times10^{-5}$a	0 ± 0a	0 ± 0a
human _ pathogens _ nosocomia	$2\times10^{-5}\pm3\times10^{-5}$a	0 ± 0a	0 ± 0a
chlorate _ reducers	$2\times10^{-5}\pm3\times10^{-5}$a	0 ± 0a	0 ± 0a
sulfite _ respiration	0 ± 0a	$3\times10^{-5}\pm1\times10^{-6}$a	$2\times10^{-5}\pm3\times10^{-5}$a
arsenate _ detoxification	0 ± 0a	0 ± 0a	$1\times10^{-5}\pm2\times10^{-5}$a
dissimilatory _ arsenate _ reduction	0 ± 0a	0 ± 0a	$1\times10^{-5}\pm2\times10^{-5}$a

注：不同小写字母表示在 0.05 水平（LSD 检验）差异显著，$n=3$。

如图 9-3 所示，12 个功能组在至少两个实验组之间具有显著差异（$P<0.05$）。GO 的暴露引起 11 个功能组的相对丰度降低（图 9-3 B~L），其中，有的与 N 元素循环相关（包括：尿素分解作用、固氮作用、硝酸盐呼吸和还原作用、氮素呼吸作用），有的与 S 元素循环相关（包括：硫酸盐和其他含硫化合物的呼吸作用），还有的与有机物降解有关（包括：化能异养作用和发酵作用）。RGO 的暴露只引起 4 个功能组的相对丰度降低（图 9-3 F、H、I 和 K），包括：硝酸盐还原作用、硫酸盐和其他含硫化合物的呼吸作用、化能异养作用。GBNs 的加入一旦引起土壤微生物菌群及其功能的改变，这些功能组所驱动的相关元素的生物地球化学循环将受到干扰[406]。另外，GO（$P<0.05$）和 RGO（$P>0.05$）的暴露都引起了 1 个功能组的相对丰度升高（图 9-3A），即，芳香族化合物降解。GBNs 分子因为包含有大量的芳香烃结构，可能会选择性富集具有芳香族化合物降解能力的菌群[218,407]。以上功能组的分析结果还说明 GO 对土壤细菌菌群功能组的影响明显强于 RGO。

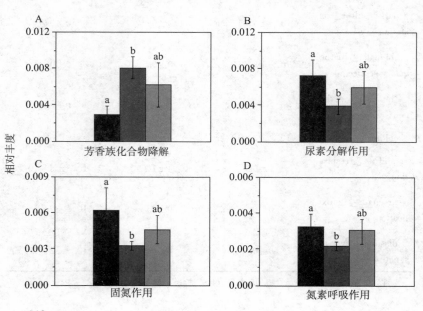

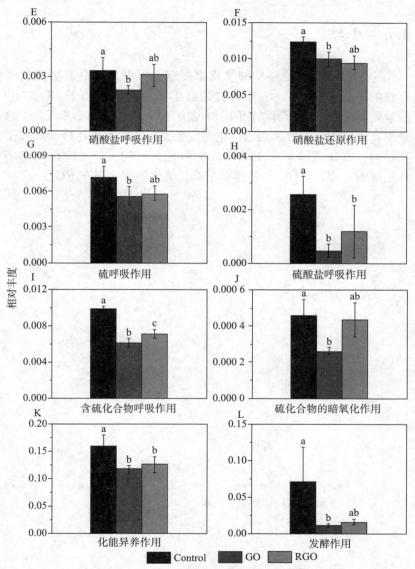

图 9-3 相对丰度具有显著差异的菌群功能组

注：误差棒代表平均值±标准差，不同小写字母表示在 0.05 水平（LSD 检验）差异显著，$n=3$。

9.6 小结

GO 和 RGO 在土壤中的暴露都引起土壤细菌菌群丰富度和多样性的降低，但 RGO 的暴露对菌群丰富度的影响强于 GO。GO 的暴露对土壤细菌菌群结构的影响强于 RGO。经过 GO 和 RGO 的暴露，土壤细菌菌群的功能性也将发生变化：芳香族化合物降解作用增强，有机质降解作用减弱，N 和 S 元素的生物地球化学循环作用减弱。而且，GO 的暴露对菌群功能性的影响强于 RGO。

第10章 氧化石墨烯对植物修复苯并[a]芘污染土壤的影响

10.1 引言

多环芳烃（PAHs）是持久性有机污染物（POPs）中的一类，被世界各国都列为优先控制的污染物[408]。作为PAHs的一种，苯并[a]芘（B[a]P）分子含有五个苯环，具有致畸性、致癌性和急性毒性[125,409]。PAHs主要来自于人类活动的释放，例如，化石燃料的不完全燃烧、石油泄漏和汽车尾气的排放等[410-412]。土壤生态系统是环境中PAHs的主要归宿[413]。PAHs对公共安全和土壤生态系统都造成严重的风险，需要从土壤中消除[414,415]。

植物修复是一种既经济又环保的消除土壤有机污染物的修复技术[416]。植物可以通过植物提取、植物代谢和增强生物降解等途径促进有机污染物的消散[417]。植物的存在可以调节微生物菌群，使其更有利于土壤污染物的生物降解[418]。据报道，已有研究通过种植紫花苜蓿和黑麦草等植物对土壤POPs进行修复[194,419,420]。

近年来，氧化石墨烯（Graphene oxide，GO）的应用和产品迅猛增长，必将导致其环境排放量的增加[400]。GO应用领域广泛，例如，净化、生物医学、吸附和催化等[421,422]。有报道称，GO产品的年产量已经达到百吨的水平[423]。GO产品所带来的GO废物将不可避免地进入土壤环境。除了废物排放，GO还可以用于环境修复和农业生产，直接施用于土壤[401]。有研究认为GO可吸附固定重金属用于土壤修复[424]。也有研究发现GO类的纳米材料含有芳香族的结构，可以增加土著PAHs降解微生物的丰富度[218,407]。

施用 GO 可能会促进 PAHs 污染土壤的生物修复[425]。He 等研究发现 GO 通过增强水传递促进蔬菜在土壤中的萌发[426]。碳纳米材料在农业领域应用的浓度已经达到较高水平[402]。

GO 产品进入土壤是不可避免的，而且数量在增加。但是，GO 对土壤污染物的消散以及修复系统的影响仍不得而知。本研究通过盆栽试验研究了植物-GO 联合作用对 B[a]P 消散以及相关土壤细菌菌群的影响。

10.2 盆栽实验方案与设计

实验用土采自山西省临汾市尧都区周边农田（36°04′N、111°34′E）的表层土（0~20cm），土壤类型属于棕壤，无 B[a]P 检出。土壤风干后，用 1mm 孔径的尼龙网筛去除石头、土块等杂质。

实验所用植物：孔雀草（*Tagetes patula* L.），菊科万寿菊属，被子植物门双子叶植物纲。作为一种对 B[a]P 污染具有很强耐性的植物[427,428]，孔雀草是从花卉植物中筛选出来的，而且具有抗逆性和适应性强等特点，在全世界各地广泛种植。这些特点无疑有利于其在污染土壤植物修复中广泛应用。花种购自北京花儿朵朵花仙子农业有限公司。

本实验所用试剂见表 10-1，GO 购自南京先丰纳米材料有限公司。

表 10-1 实验试剂

试剂名称	生产厂家	备注
B[a]P	美国百灵威公司	纯度 96%
十氟联苯	美国百灵威公司	纯度 99%
正己烷	天津市康科德科技有限公司	色谱纯、分析纯
丙酮	天津市康科德科技有限公司	分析纯
硅镁型吸附剂	上海安谱实验科技股份有限公司	0.15~0.55mm
30%过氧化氢	天津市福晨化学试剂厂	分析纯
无水乙醇	天津市福晨化学试剂厂	分析纯

第10章 氧化石墨烯对植物修复苯并[a]芘污染土壤的影响

10.2.1 盆栽土的制备

B[a]P标准品经正己烷溶解后加入土壤。有机溶剂在排气室挥发72h后与土壤充分混匀，最终浓度为10mg/kg干土。污染土在25℃温室中平衡4周后装盆，每盆1.5kg土。按照参考文献，将提前准备好的GO悬浊液与花盆中的土样充分搅拌混合[172]。

10.2.2 盆栽实验设计

孔雀草种子催芽和育苗步骤同第2章。待孔雀草幼苗长出4片真叶（约高8cm）移植到经平衡后装有染毒土的花盆中，并转移到温室中培养。

共设置6组处理："Control"代表没有GO添加，也没有植物种植；"GO-0-plant"代表种植植物，但没有GO添加；"GO-100"代表以100mg/kg浓度添加GO，但没有种植植物；"GO-10-plant"代表种植植物，而且以10mg/kg浓度添加GO；"GO-50-plant"代表种植植物，而且以50mg/kg浓度添加GO；"GO-100-plant"代表种植植物，而且以100mg/kg浓度添加GO。

田间管理：温室维持日间温度为25℃，夜间温度为20℃，土壤水分维持在田间持水量的50%，并适度通风，待植物成熟后收获，生长时间为90d。培养结束后，收集土壤和植物样品，贮藏于-20℃，并尽快分析测定。

10.2.3 土壤和植物B[a]P的测定方法

土壤和植物样品经冷冻干燥，粉碎（土样需再过0.15mm筛）后，定量称取样品并加入代标十氟联苯（Decafluorobiphenyl），用滤纸包裹后置于平底烧瓶中，加入200mL提取溶剂（丙酮：正己烷=1:1），平底烧瓶放在水浴锅中，连续抽提24h。提取液在旋转蒸发仪上浓缩至1~2mL，而后溶剂置换成正己烷，接着继续浓缩至1mL，完成溶剂替换。提取液经过硅镁吸附剂（Florisil）层析柱净化后，用150mL正己烷洗脱，洗脱液用轻柔的氮气吹至近干，用色谱纯正己烷定容至 1.0mL，上机（GC-MS）测定分析。

GC-MS 参数：分离用的是毛细管柱（J&W DB-5MS，30m×0.32m×0.25μm）。载气为氮气，流速：2mL/min。分离柱升温程序：起始温度 90℃（保持 1min），然后以 15℃/min 的速度升至 180℃（保持 1min），以 10℃/min 升到 285℃（保持 1min）；最后以 7℃/min 升到 290℃（保持 2min），无分流进样（1μL）。土壤和植物样品的代标回收率分别为 78%～96% 和 75%～103%。

10.2.4 土壤细菌菌群分析方法

主要步骤包括：DNA 提取、PCR 反应、高通量测序、序列加工处理和生物信息学分析。通过 PCR 反应扩增细菌 16S rRNA V3～V4 区的 DNA 片段，PCR 扩增引物：338F 和 806R。原始序列已经上传至 NCBI Sequence Read Archive（SRA）数据库（Accession Number：PRJNA549382）。

α 多样性分析包括：Chao1 指数、ACE 指数和 Shannon 指数。β 多样性分析通过 NMDS 分析研究样品间的菌群结构变化。通过 UniFrac 举例度量，并通过 NMDS 分析图呈现。各菌群在菌门水平的丰富度变化通过小提琴图（violin plots）呈现。

10.2.5 数据分析

所有实验都进行 3 次重复，所得的数据均进行方差分析，文中误差棒代表了标准差，结果以 SPSS 19.0 统计软件进行分析，数据的显著水平均指 $P<0.05$（ANOVA，Tukey）。

10.3 土壤 B[a]P 的消散以及植物对 B[a]P 的提取

不同处理条件下的土壤 B[a]P 消散率，见表 10-2。B[a]P 消散率由高到低分别为："GO-0-plant"（98.30%）＞"GO-100-plant"（96.37%）＞"Control"（93.48%）＞"GO-100"（89.61%）。与没有

第 10 章 氧化石墨烯对植物修复苯并 [a] 芘污染土壤的影响

种植孔雀草的实验组相比,种植植物显著($P<0.05$)提高了土壤 B [a] P 的消散率。与没有添加 GO 的实验组相比,GO 的添加显著($P<0.05$)抑制了土壤 B [a] P 的消散率。"GO-100-plant"实验组的 B [a] P 消散率显著($P<0.05$)高于"GO-100"实验组,而显著($P<0.05$)低于"GO-0-plant"实验组,说明种植植物减弱了 GO 对 B [a] P 消散率的抑制作用。

表 10-2 不同处理条件下土壤 B [a] P 消散率(%)

处理	B [a] P 消散率
Control	93.48±0.50a
GO-0-plant	98.30±0.88b
GO-100	89.61±1.00c
GO-100-plant	96.37±0.63b

注:不同小写字母表示在 0.05 水平(Tukey 检验)差异显著,$n=3$。

不同 GO 添加浓度对土壤 B [a] P 消散率和植物提取 B [a] P 的影响,见表 10-3。GO 的添加浓度升高,土壤 B [a] P 消散率降低,植物体内 B [a] P 的提取量也减少。"GO-100-plant"实验组的 B [a] P 消散率显著($P<0.05$)低于"GO-0-plant"实验组。"GO-10-plant"、"GO-50-plant"和"GO-100-plant"实验组的植物地上部 B [a] P 的浓度均显著($P<0.05$)低于"GO-0-plant"实验组。"GO-100-plant"实验组的植物根部 B [a] P 的浓度显著($P<0.05$)低于"GO-0-plant"实验组。如图 10-1 所示,植物的地上部和地下部生物量并没有因 GO 的添加而受到显著影响($P>0.05$)。

表 10-3 不同 GO 浓度处理条件下土壤 B [a] P 消散率和植物体 B [a] P 的浓度

处理	B [a] P 消散率(%)	植物地上部 B [a] P 浓度(mg/kg)	植物根部 B [a] P 浓度(mg/kg)
GO-0-plant	98.30±0.88a	0.14±0.02a	1.03±0.13a
GO-10-plant	97.80±0.58ab	0.09±0.01b	0.81±0.17ab
GO-50-plant	97.01±0.51ab	0.08±0.02b	0.80±0.12ab
GO-100-plant	96.37±0.63b	0.05±0.01b	0.58±0.13b

注:不同小写字母表示在 0.05 水平(Tukey 检验)差异显著,$n=3$。

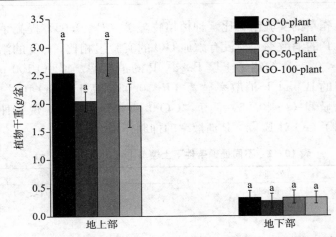

图 10-1 不同 GO 浓度处理 B [a] P 污染土壤条件下植物地上部和地下部干重

注：误差棒代表平均值±标准差，相同小写字母表示在 0.05 水平（Tukey 检验）差异不显著，$n=3$。

由表 10-4 可见，植物对土壤中 B [a] P 的提取和转移能力非常弱，其富集系数和转移系数分别低于 1.4 和 0.14。植物积累的 B [a] P 含量占因种植植物而提高的消散量的比重不到 0.18%。B [a] P 属于 POPs 的一种，POPs 和土壤颗粒具有很强的结合力，很难被植物组织所富集[194]。

表 10-4 孔雀草对 B [a] P 的富集系数 (BF)、转移系数 (TF) 和植物富集对 B [a] P 消散的贡献率

处理	BF	TF	贡献率（%）
GO-0-plant	1.372±0.170a	0.137±0.003a	0.14±0.03a
GO-10-plant	0.882±0.061b	0.115±0.042a	0.09±0.01a
GO-50-plant	0.665±0.097bc	0.104±0.024a	0.18±0.06a
GO-100-plant	0.439±0.089c	0.085±0.020a	0.17±0.07a

注：不同小写字母表示在 0.05 水平（Tukey 检验）差异显著，$n=3$。富集系数 (BF)＝植物体内 B [a] P 浓度/土壤 B [a] P 浓度；转移系数 (TF)＝植物地上部 B [a] P 浓度/植物地下部 B [a] P 浓度；植物贡献率＝（植物体内 B [a] P 浓度/土壤 B [a] P 增多的消散量）×100%。

10.4　土壤细菌菌群测序结果

经 DNA 提取、PCR 反应和高通量测序技术测序后发现，12 个样品（4 组）共获得 549 366 个高质量序列，序列长度主要分布在 423~461bp。从表 10-5 可见，"Control"、"GO-0-plant"、"GO-100" 和 "GO-100-plant" 四个实验组分别获得 40 336 个、54 660 个、41 129 个和 46 997 个序列。

表 10-5　土壤细菌菌群的丰富度和多样性

	Sequences	3% distance		
		ACE	Chao1	Shannon
Control	40 336±3 634a	3 353±854ab	3 249±829ab	10.56±0.03ab
GO-0-plant	54 660±7 591b	4 557±486b	4 394±583b	10.68±0.06b
GO-100	41 129±2 614a	2 750±345a	2 686±247a	10.58±0.09ab
GO-100-plant	46 997±4 626ab	2 989±255a	2 878±160a	10.36±0.20a

注：不同小写字母表示在 0.05 水平（Tukey 检验）差异显著，$n=3$。

10.5　土壤细菌菌群 α-多样性分析结果

基于 97% 的相似性和 Greengenes 数据库的比对结果，将各序列归入不同的分类水平（门、纲、目、科和属）的 OTU，见表 10-6。在 5 个分类水平上，各实验组的 OTU 数由高到低依次为："GO-0-plant" > "GO-100-plant" > "Control" > "GO-100"。所有样品的稀释曲线（图 10-2）都趋于饱和平稳状态，说明测序深度足以覆盖所有的细菌菌群。

ACE 和 Chao1 指数反映微生物菌群的种群丰富度，如表 10-5、图 10-2 所示，细菌种群丰富度由高到低依次为："GO-0-plant" > "Control" > "GO-100-plant" > "GO-100"。与对照相比，种植植物增加了细菌菌群丰富度（$P>0.05$），而添加 100mg/kg GO 时降低了菌群丰富度（$P<0.05$）。"GO-100-plant"实验组的菌群丰富度显著（$P<0.05$）低于"GO-0-plant"实验组，说明在种植孔雀草的B[a]P污染土壤中，添加GO引起了细菌菌群丰富度的下降。

表 10 - 6　不同分类水平的 OTU 数

分类水平	Control	GO-0-plant	GO-100	GO-100-plant
门	2 665±299a	3 033±278a	2 538±177a	2 822±132a
纲	2 621±290a	2 970±277a	2 490±176a	2 762±127a
目	2 234±265a	2 494±263a	2 126±149a	2 391±133a
科	1 905±270a	2 162±233a	1 807±147a	2 078±77a
属	1 013±181a	1 216±108a	946±56a	1 127±71a

注：不同小写字母表示在 0.05 水平（Tukey 检验）差异显著，$n=3$。

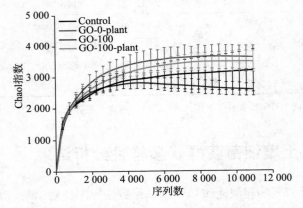

图 10 - 2　细菌菌群 Chao1 指数的稀释曲线
注：误差棒代表平均值±标准差。

Shannon 指数反映菌群多样性，如表 10 - 5 所示，细菌菌群多样性由高到低依次为："GO-0-plant" > "GO-100" > "Control" > "GO-100-plant"。与对照相比，种植植物或者添加 GO（100mg/kg）都增加了菌群多样性（$P>0.05$）。"GO-100-plant"实验组的 Shannon 指数显著（$P<0.05$）低于"GO-0-plant"实验组，说明在种植孔雀草的 B[a]P 污染土壤中，添加 100mg/kg 的 GO 明显引起了细菌菌群丰富度的下降。

10.6 土壤细菌菌群结构分析结果

根据系统分类学结果，4 个实验组的优势菌群在门水平上是一致的，共鉴定出 12 个门，如图 10-3 所示，包括：Acidobacteria、Actinobacteria、Armatimonadetes、Chlamydiae、Chloroflexi、Cyanobacteria、Firmicutes、Ignavibacteriae、Latescibacteria、Proteobacteria、Spirochaetae 和 Deinococcus-Thermus。这些共有的菌群分别占到"Control"、"GO-0-plant"、"GO-100"和"GO-100-plant"实验组的 98.3%、97.7%、97.8% 和 97.5%。由此可见，在门水平上，植物和 GO 单独或共同存在都没有明显改变 B [a] P

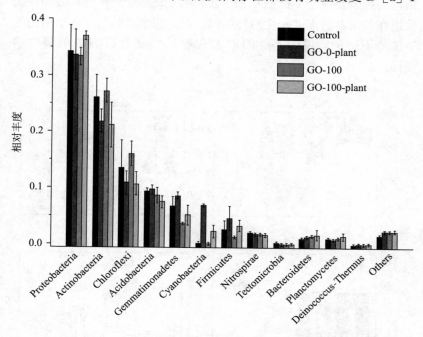

图 10-3 土壤优势细菌菌群在门水平的相对丰度

注：误差棒代表平均值±标准差。"Others"代表此图中不能归入任何已知菌门的菌群。

污染土壤原有门水平的优势菌群。但是，不同实验组在各菌门的相对丰度是不同的。

如图10-4所示,"GO-0-plant"和"GO-100-plant"实验组在各菌群的丰度变化趋势一致。"GO-0-plant"和"GO-100-plant"实验组在7个门的丰度都高于"Control"和"GO-100"实验组，包括：Armatimonadetes、Chlamydiae、Cyanobacteria、Firmicutes、Ignavibacteriae、Spirochaetae和Deinococcus-Thermus。而且,"GO-0-plant"和"GO-100-plant"实验组在2个门的丰度都低于"Control"和"GO-100"实验组，包括：Actinobacteria和Chloroflexi。以上结果说明，在B[a]P污染土壤种植植物对细菌菌群的影响要强于GO的添加。虽然"Control"和"GO-100"实验组在各菌群的丰度变化趋势相似，但是两个实验组在各个菌门的丰度是不同的，说明GO的添加在较低程度上影响细菌菌群的结构。

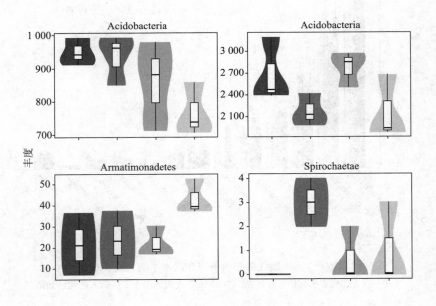

第 10 章 氧化石墨烯对植物修复苯并 [a] 芘污染土壤的影响

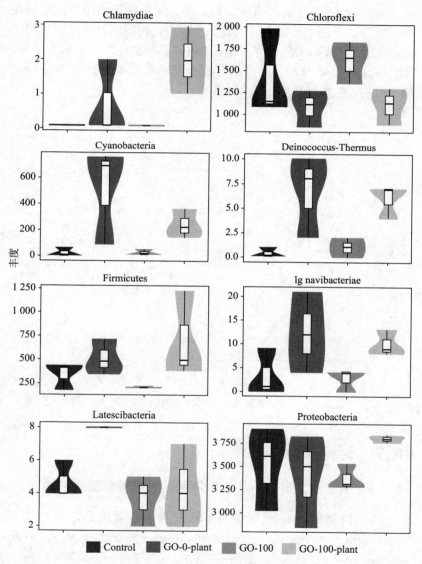

图 10-4 Violin 图描绘已知 12 个门的丰度（Taxa Abundance）和分布密度

如图 10-5 所示，NMDS 分析证实了 4 个实验组的菌群结构差异。在 NMDS 图中，每个点代表 1 个样本，相同颜色的点属于相同的实验组。NMDS 分析结果显示，4 组样品明显地趋向于两个集合，"GO-0-plant"和"GO-100-plant"实验组处在一个集合，"Control"和"GO-100"实验组处于另一个集合。

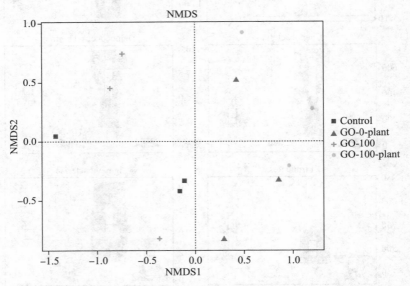

图 10-5 不同处理条件下细菌菌群的 NMDS 分析图

10.7 讨论

10.7.1 植物对土壤 B[a]P 消散和相关细菌菌群的影响

种植植物能促进土壤 POPs 的去除。有报道称，种植南瓜 (*Cucurbita pepo* ssp.)[429]、*Brassica nigra*[430]、黑麦草 (*Lolium perenne* L)[431]、苜蓿 (*Medicago sativa* L.) 和高羊茅 (*Festuca arundinacea* Schreb)[186]均可以增强土壤 PCBs 的消散。就 PAHs 而言，研究发现，种植玉米 (*Zea mays* L.)[432]和黑麦草[433]分别

第10章 氧化石墨烯对植物修复苯并[a]芘污染土壤的影响

可以增强土壤芘和菲的消散。

本研究发现，种植孔雀草可以增强土壤B[a]P的消散。土壤有机污染物消散的机理主要包括：植物积累、植物挥发、光降解和生物降解[432]。在本研究中，由于种植植物处理和对照处理的非生物损失是相同的，植物对土壤B[a]P消散的增强作用主要归因于生物因素。在本研究中，孔雀草对B[a]P的积累作用很弱，其富集系数和转移系数分别低于1.4和0.14。植物积累的B[a]P含量仅占因植物存在而增加的B[a]P消散量的0.18%。POPs的强疏水性导致其在土壤中易于结合在土壤颗粒而不易被植物组织提取[194]。因此，植物激发的微生物降解作用可能在植物对土壤B[a]P消散的促进作用中起主要作用。植物可以释放根系分泌物，提高微生物的活性和调整微生物菌群的结构，进而促进POPs的生物降解[434]。Yoshitomi等研究发现 *Zea mays* L. 的根系分泌物促进土壤微生物的生长，从而增强PAHs的降解[435]。本研究中，种植孔雀草导致土壤细菌菌群丰富度和多样性增加，菌群结构也发生变化。这些菌群变化可能在植物促进土壤B[a]P消散过程中发挥重要的作用。

10.7.2 GO对土壤B[a]P消散和相关细菌菌群的影响

GO对POPs具有很强的吸附亲和性[436]，导致POPs从土壤中消散量减少。在本研究中，GO（100mg/kg）的添加导致土壤B[a]P的消散明显减少。Ren等认为碳纳米材料（CNMs）均具有很强的吸附能力，会导致土壤中有机污染物的生物可利用性降低[437]。目前，很多研究认为CNMs具有抑菌活性。CNMs抑菌机理主要包括：破坏细胞壁、细胞膜和线粒体，造成氧化胁迫和代谢紊乱[348,438-440]。但是，由于土壤微生物菌群的复杂性，研究人员关于CNMs在土壤基质中对微生物的影响还没有统一的认识[425]。Tong等认为C_{60}导致微生物生物量减少[278]。Rong等研究发现石墨烯或者氧化石墨烯在土壤中暴露引起了土壤微生物菌群丰富度和多样性的升高[441]。Khodakovskaya等却发现土壤微生物菌群的丰富度和多样性不会受到 CNMs 的影响[442]。还有研究人员认为 CNMs 对土壤微生物菌群的影响因微生物的种类而不同，对 CNMs 具有耐性的微生物种群将会得到富集[382]。Mu 等研究发现氧化石墨烯量子点（GOQDs）的添加提高了 PAHs 降解菌的增殖，可能是因为 GOQDs 分子结构中含有很多芳香烃结构[407]。在本研究中，GO（100mg/kg）添加后，土壤细菌菌群丰富度和多样性升高了，但是不显著。此外，GO 对细菌在门水平对优势菌群影响很小。

10.7.3　GO 和植物对土壤 B［a］P 消散和细菌菌群的交互影响

当 GO 和植物共存时，GO 对植物生长的影响同时也影响土壤 POPs 的去除效率。在水培条件下，CNMs 抑制植物生长，表现在抑制光合作用、幼苗生长和生物量等方面[443]。但是，土壤培养和水体培养不同，GO 在土壤基质中的运输受限。而且，在土壤中，植物根和 GO 的接触减少会导致 GO 的植物毒性大幅度减弱[444]。此外，还有研究人员认为，在土壤环境中，GO 可以促进水分传输而促进植物生长[426]。本研究发现 GO 在低于 100mg/kg 浓度下对孔雀草的生物量没有明显的影响。一般认为，CNMs 吸附有机污染物，也会导致植物对有机污染物的生物积累减少。Hamdi 等研究发现 CNMs 在土壤中降低杀虫剂的生物可利用性，导致其在莴苣中的积累量减少[445]。在本研究中，土壤中 GO 的添加量增加，孔雀草对 B［a］P 的积累量减少。

当 CNMs 和植物在污染土壤中共存时，污染物的消散不仅与 CNMs 对消散的抑制作用有关，还与植物对消散的促进作用有关。本研究发现种植孔雀草减弱了 GO 对 B［a］P 消散的抑制作用。植物分泌根系分泌物，促进酸溶性污染物的解析，促进污染物的生物降解[446]。CNMs 对污染物的吸附作用可能被根系分泌物的解析作用所掩盖。

植物只能吸收土壤中溶解态污染物，而微生物可以利用溶解态和吸附态的污染物[437]。因此，在本研究中，"GO-100-plant" 处理的消散率介于 "GO-100" 处理和 "GO-0-plant"，这个结果依赖于

第 10 章　氧化石墨烯对植物修复苯并 [a] 芘污染土壤的影响

孔雀草和 GO 的综合作用。同样，GO 和植物的综合作用导致了本研究中土壤细菌菌群的变化。在本研究中，种植植物对细菌菌群的影响大于 GO 的添加。

10.8　小结

种植植物（孔雀草）显著（$P<0.05$）增强了土壤 B [a] P 的消散。GO（100mg/kg）的添加显著（$P<0.05$）减少了土壤 B [a] P 的消散。当植物和 GO 共同存在时，植物减弱了 GO 对土壤 B [a] P 消散的抑制作用。与对照相比，GO 的添加或者种植植物对 B [a] P 污染土壤的细菌菌群多样性和丰富度均没有显著的影响（$P>0.05$）。与只种植植物的处理相比，植物和 GO 联合作用显著（$P<0.05$）降低了菌群丰富度和多样性。种植植物对细菌菌群结构的影响强于 GO 的影响。植物的存在减弱了 GO 对土壤 B [a] P 消散的抑制作用。

第 11 章　碳基负载纳米材料对植物修复复合污染初探

11.1　引言

近年来，土壤污染趋向严重化和复杂化，有研究报道 64% 的污染场地属于有机污染物和无机污染物的复合污染[32]。例如，电子垃圾成分复杂，把 Pb、Hg、Cr、Cd 和 Ni 等几十种重金属和聚氯乙烯、多氯联苯（PCBs）、多溴联苯等多种有机污染物同时带入土壤。重金属 Cd 和 PCBs 是土壤环境中两类重要污染物，往往先后或同时进入土壤环境形成复合污染。

目前，污染土壤的修复方法主要包括植物修复、微生物修复、物理修复和化学修复。植物修复技术能有效地去除重金属和有机污染物[96]，而且有投资少、不破坏场地结构、绿化环境等优点。用于植物修复技术的植物应该是一类具有一定特殊功能的植物，包括重金属超累积植物、某些藻类及突变株和对有机污染物具有一定特异降解功能的植物，在选用这些植物时可以根据环境治理的目标而有所不同或侧重[2]。然而，植物修复技术也会因周期长、深层土壤修复能力弱等因素降低修复效率，有必要采取强化措施来提高植物修复的效率。

纳米材料具有吸附、催化、辐射、吸收等特性，由于其大量的微界面及微孔性，可以强化各种界面反应，如对重金属的表面及专性吸附反应等，在重金属污染和持久性有机污染物污染土壤治理中将发挥显著作用[447]。Nanoscale zero-valent iron（nZVI）具有比表面积大、还原性强等特点，能有效除去环境中的有机污染物和重金属[448]。nZVI对氯代有机污染物具有还原脱氯的作用，能高效去除环境中的PCBs[449]。氧化石墨烯（Graphene oxide，GO）拥有大量的含氧官能团和高表面积，吸附土壤中的有机和无机污染物能力强。生物炭（Biochar）是由生物残体在缺氧的情况下，经高温慢热解（通常<700℃）产生的一类稳定的、难熔的、高度芳香化的、富含碳素的固态物质[450]。生物炭可以降低土壤中污染物的有效性，表现出修复污染土壤和促进作物生长的双重效果。nZVI虽然可应用于降解PCBs，但其易被空气中的氧气氧化并且易团聚[451,452]。所以，为了减少nZVI氧化和团聚，可将nZVI负载到其他载体上，保持nZVI固有还原特性的基础上增强其稳定性；如果载体也有助于污染物的去除，可以结合两者优点，共同为污染物修复服务。

纳米材料的制备方法包括化学还原法、化学沉淀法、溶胶凝胶法、水热法和热分解法等[453]。本研究采用液相还原法制备nZVI，液相还原法反应条件温和，易分离产物，反应速度快，操作简单方便，而且成本低[454]。本研究分别将nZVI负载到GO和Biochar，实现以上设想[455]。石墨烯负载纳米铁（RGO-nZVI）和生物炭负载纳米铁（Biochar-nZVI）通过碳材料片层对铁颗粒的间隔作用，可避免或降低纳米材料的团聚，同时减少nZVI在环境中的氧化。

本研究针对Cd-PCBs复合污染土壤，选择具有Cd超积累特性的植物孔雀草作为修复植物，制备纳米强化剂并进行强化植物修复的盆栽试验，旨在通过观察植物生长状况对纳米材料强化植物修复初步判断其可行性，以期为Cd-PCBs复合污染土壤的植物修复探寻友好型强化剂。

11.2 盆栽实验方案与设计

本实验所特用试剂见表11-1。

表 11-1　实验试剂

试剂名称	生产厂家	纯度
氢氧化钠	天津市元立化工有限公司	分析纯
聚乙二醇 4000	北京鼎国昌盛生物技术有限公司	分析纯
硫酸亚铁	天津博迪化工股份有限公司	分析纯
硼氰化钾	天津市科密欧化学试剂有限公司	分析纯
氧化石墨烯（GO）	苏州奥索特新材料有限公司	纯度＞99%

本研究所选植物是孔雀草（*Tagetes patula* L.），试验用土采自天津市北辰区红光农场（39°11′N、117°01′E）表层土（0~20cm），所采土壤为潮土，其理化性质见前文。

11.2.1　nZVI 及负载材料的制备和表征方法

1. nZVI 的制备

反应方程式如下：

$$Fe^{2+} + 2BH_4^- + 6H_2O \longrightarrow Fe + 2B(OH)_3 + 7H_2$$

在氮气保护下，将一定量的 $FeSO_4 \cdot 7H_2O$ 和聚乙二醇溶于水中，再与无水乙醇按比例（无水乙醇：水=4:1）混合后倒入反应装置（图 11-1）的三口烧瓶中，在搅拌器开启条件下，将移入恒压漏斗中的 1mol/L KBH_4 溶液逐滴加入三口烧瓶，继续搅拌反应 40min。磁铁分离沉淀后，分别用去离子水和无水乙醇洗涤 3 次，最后在 −55℃冷冻干燥，制得 nZVI。

2. RGO-nZVI 的制备　GO，纯度＞99%，根据产品说明，GO 通过改良的 Hummers[456] 方法制备而得。将一定质量的 GO 置于三口烧瓶中，其余步骤与 nZVI 制备过程一致，控制 $FeSO_4 \cdot 7H_2O$ 加入量与 nZVI 制备过程保持一致，nZVI 在 GO 表面上原位生长，最终制得 RGO-nZVI。

3. 生物炭的制备　以干燥木屑为原材料，在高温（600℃）缺氧状态下在马弗炉中制备而成。

4. Biochar-nZVI 的制备　参考上述 RGO-nZVI 制备的方法，将 GO 换成 Biochar 而得。

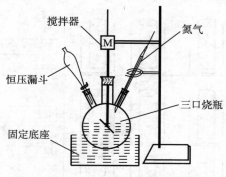

图 11-1 nZVI 制备反应装置

5. 纳米材料的扫描电子显微镜（SEM）表征 样品粉末分散于无水乙醇中，超声 10min，进行 SEM 测试（LEO-1530VP，Germany）。

11.2.2 盆栽实验

1. 实验设计 根据图 11-2，纳米强化剂的添加浓度分别为：T1、T2 和 T3 均为 500mg/kg 干土（0.05%）；T4 为 5 000mg/kg（0.5%）。每个处理 3 次重复，均在相应位置按浓度混合加入纳米材料，植物生长时间为 90d。

2. 染毒处理 Cd 投加浓度为 10mg/kg，投加的形态为分析纯试剂 $CdCl_2 \cdot 2.5H_2O$，以固态加入土壤中；PCBs 选择 Aroclor1242（AccuStandard Inc.，99%），经有机溶剂溶解后加入土壤中；污染物与土壤充分混匀，平衡 4 周后装盆，每盆 1.5kg 土，3 次重复。

3. 盆栽试验详细步骤同第 2 章。

4. Cd 和 PCBs 测定 土壤和植物样品 Cd 含量的测定步骤同第 2 章；土壤和植物样品 PCBs 含量的步骤参考第 3 章。

在测定 Aroclor1242 进行定性和定量时，单个目标化合物定性利用多氯联苯混合标样结合质谱分析定性。不同氯取代 PCBs 的定性、定量离子如表 11-2 所示，参考文献[457]进行定量分析。

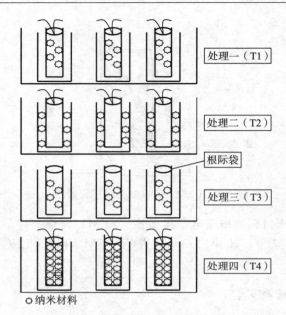

图 11-2 强化剂对植物修复影响的实验设计图

表 11-2 不同氯取代 PCBs 的定性和定量离子

组分	起始采集时间（min）	终点采集时间（min）	定量离子	定性离子
一氯联苯	14.0	17.0	188	152 188 192
二氯联苯	17.0	20.5	222	222 224 226
三氯联苯	20.5	25.5	256	223 256 260
四氯联苯	25.5	31.8	292	290 257 259
五氯联苯	31.8	35.2	326	328 324 330

11.2.3 数据处理

图表制作采用 Origin8.5，所有实验都进行 3 次重复，如有特殊情况，会在文中图表标注说明。误差棒代表了标准差，结果以 SPSS 19 统计软件进行分析。

11.3 纳米强化剂的 SEM 表征结果

本研究准备了 GO、nZVI、RGO-nZVI、Biochar、Biochar-nZVI 5 种纳米强化剂,其 SEM 结果如图 11-3 所示。GO 呈片层结构。nZVI 呈颗粒状,由于其粒度小,表面能大且具有磁性,所以

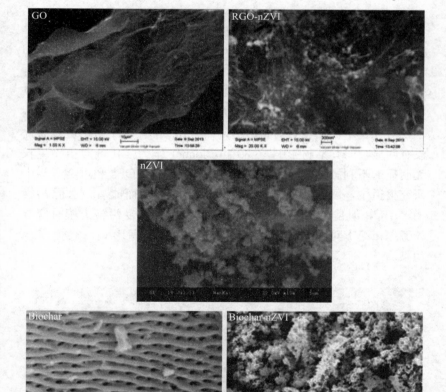

图 11-3 纳米强化剂的 SEM 结果

出现团聚[458]。RGO-nZVI 有效减少 nZVI 的团聚,使 nZVI 散布在 RGO 片层表面或片层间。Biochar 呈致密网状结构,尺度超过纳米级。Biochar-nZVI 是把 nZVI 负载到 Biochar 表面,从 SEM 结果可看出,大量 nZVI 堆积在生物炭表面并发生团聚。

11.4 纳米材料对孔雀草生长状况的影响

如图 11-4 所示,观察发现 500mg/kg GO、RGO-nZVI、Biochar 和 Biochar-nZVI 没有造成植物枯萎;500mg/kg nZVI 使个别植物枯萎;5 000mg/kg GO 和 Biochar 没有造成植物枯萎;5 000mg/kg nZVI、Biochar-nZVI 和 RGO-nZVI 处理组抑制孔雀草生长,叶片陆续发黄,并在 30d 内枯萎。以上结果说明,高浓度纳米铁对孔雀草生长具有不利影响,而且用 GO 和 Biochar 负载后并没有明显减弱其不利影响。GO 和 Biochar 作为强化剂对植物生长无明显的不利影响。尽管 nZVI 在修复土壤的 POPs 和重金属污染研究报道中,展现出强还原能力和强吸附能力而能有效去除污染物[448,449],但是在利用 nZVI 修复或强化植物修复污染土壤时应首先考虑 nZVI 对植物生长以及生态环境的毒害影响,避免二次污染。

图 11-4 纳米材料对孔雀草生长状况的影响

通过观察还发现,500mg/kg nZVI 处理的盆栽试验中,在根际袋内和根际袋外添加纳米材料都会导致根际袋内的植物最终枯萎,而且将在洁净土正常生长的植物移栽到 5 000mg/kg nZVI 处理的花盆中任何位置都会逐渐枯萎,说明纳米材料在本研究土壤介质(潮土)中具有快速的迁移能力。在应用纳米材料原位修复污染土壤时,纳米材料可以迅速迁移至附近的污染区,同时也需警惕纳米修复材料的迁移污染。

11.5 纳米强化剂对孔雀草生物量的影响

5 种强化剂对孔雀草平均生物量的影响,如图 11-5 所示。结合孔雀草株高和地上部鲜重可以发现,GO 作为强化剂时提高了植物生物量;低浓度 RGO-nZVI 作为强化剂时提高了植物生物量,高浓度 RGO-nZVI 作为强化剂时抑制植物生长;nZVI、Biochar 和 Biochar-nZVI 作为强化剂时,低浓度没有明显提高植物生物量,高浓度明显抑制植物生长。

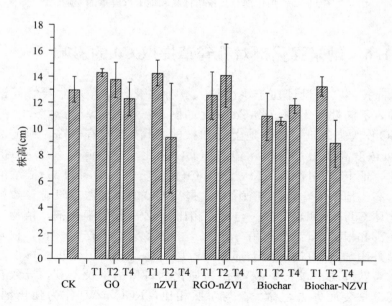

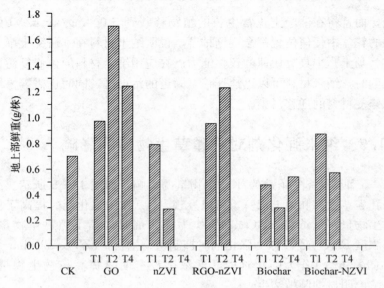

图 11-5 强化剂对孔雀草株高和地上部鲜重的影响

11.6 纳米强化剂对孔雀草提取 Cd 的影响

按照实验设计在土壤中加入 5 种强化剂，经过盆栽修复试验，不同处理组的植物 Cd 浓度如图 11-6 所示。与对照相比，GO 作为强化剂，植物地上部和根的 Cd 浓度均有升高，地上部 Cd 浓度高于根；与根际袋添加 GO 相比，在根际袋外添加 GO 时，地上部 Cd 浓度下降，根部 Cd 浓度升高，并且根部 Cd 浓度远高于地上部，转移系数降低；提高 GO 的添加浓度没有导致植物体 Cd 浓度明显升高。与对照相比，nZVI 作为强化剂，植物地上部和根的 Cd 浓度均稍有升高；与根际袋添加 nZVI 相比，根际袋外添加 nZVI 时，地上部 Cd 浓度升高，根部 Cd 浓度下降，转移系数升高；高浓度 nZVI 作为强化剂对植物生长不利，导致孔雀草枯萎而失去生命力。与对照相比，RGO-nZVI 作为强化剂，植物体 Cd 浓度变化趋势与 GO 相似，只是高浓度 RGO-nZVI 作为强化剂导致孔雀草枯死。与对照相比，Biochar 作为强化剂，植物地上部和根的 Cd 浓度均有降低；与根际袋添加 Biochar 相比，根际袋外添加 Biochar 时，地上部 Cd 浓度稍有升高；提高 Biochar 的添加浓度时，植物体 Cd 浓度也稍有升高，总体而言，Biochar 作为强化剂没有明显提高植物体 Cd 浓度。与对照相比，Biochar-nZVI 作为强化剂，植物体 Cd 浓度变化趋势与 Biochar 相似，只是植物体 Cd 浓度比 Biochar 作为强化剂时略有升高；而且与 nZVI 和 RGO-nZVI 作为强化剂时的效果一样，高浓度 RGO-nZVI 导致孔雀草枯萎而失去生命力。

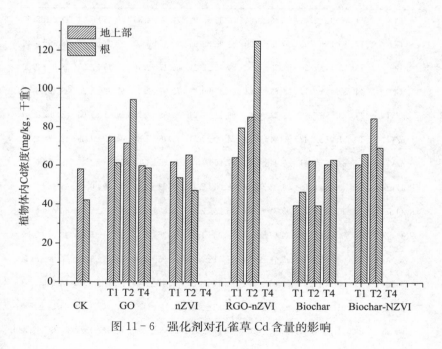

图 11-6　强化剂对孔雀草 Cd 含量的影响

11.7　纳米强化剂对土壤 PCBs 去除率的影响

以 GO、nZVI、RGO-nZVI、Biochar 和 Biochar-nZVI 5 种强

化剂，按照实验设计在土壤中加入并经过盆栽试验，土壤 PCBs 去除率如表 11-3 所示。GO 作为强化剂，土壤中低氯联苯（1~3 Cl）去除率维持在 85% 以上；四氯联苯和五氯联苯去除率为 65%~75%；六氯联苯去除率在 50% 以下。GO 作为强化剂，在根际袋和非根际袋添加 GO，PCBs 去除率差异不明显；与种植物相比，不种植物时土壤 PCBs 去除率略有下降，其中六氯联苯下降幅度最大；提高 GO 的添加浓度没有提高 PCBs 的去除率。

表 11-3　5 种强化剂处理对土壤 PCBs 去除率的影响

处理		1Cl	2Cl	3Cl	4Cl	5Cl	6Cl
GO	T1	85.02±2.52	96.46±0.84	86.90±3.92	74.49±4.35	66.45±5.36	49.21±5.55
GO	T2	86.14±3.92	96.63±1.90	86.95±5.05	74.05±6.00	67.27±3.64	48.36±6.46
GO	T3	83.48±1.26	95.63±0.57	84.13±2.21	69.47±2.48	62.76±2.33	38.31±8.32
GO	T4	82.94±2.14	96.63±0.64	86.60±0.80	72.97±2.23	67.32±5.20	48.47±5.48
nZVI	T1	83.58±2.38	95.87±0.61	85.34±2.51	71.58±4.70	64.81±5.06	42.50±12.44
nZVI	T2	81.07±4.48	95.24±0.63	84.00±2.51	70.16±3.67	64.55±2.69	42.52±2.54
nZVI	T4	84.75±0.61	96.44±0.44	87.90±1.02	75.56±1.83	68.68±2.05	49.98±6.77
RGO-nZVI	T1	40.30±1.69	96.33±0.48	88.05±2.19	74.99±2.25	67.79±0.81	44.46±3.41
RGO-nZVI	T2	46.40±1.77	95.74±0.78	86.28±2.85	73.86±4.59	67.48±5.50	43.58±5.07
RGO-nZVI	T4	73.42±17.69	96.66±0.14	88.98±1.89	76.30±4.67	67.81±5.63	43.31±9.80
Biochar	T1	46.18±70.61	95.78±0.80	87.71±0.76	76.96±0.94	71.57±1.85	54.18±10.59
Biochar	T2	62.97±2.55	92.85±3.94	81.14±8.04	67.42±8.44	44.08±37.75	36.03±8.83
Biochar	T3	81.85±4.66	93.68±4.54	83.82±9.81	73.16±11.82	68.96±1.60	44.34±1.90
Biochar	T4	81.35±5.77	93.10±4.24	81.07±9.95	66.88±13.43	62.18±1.04	39.40±13.88
Biochar-nZVI	T1	82.99±1.02	96.03±0.56	87.58±1.27	75.69±2.46	69.59±4.16	46.73±3.87
Biochar-nZVI	T2	82.71±5.19	95.50±1.13	87.72±1.30	76.29±1.68	71.30±3.32	53.74±9.98
Biochar-nZVI	T4	80.52±1.90	94.22±1.67	84.45±5.04	71.64±6.50	66.73±5.27	44.62±3.24

nZVI作为强化剂,土壤中低氯联苯(1~3 Cl)去除率维持在83%以上;四氯联苯和五氯联苯去除率为65%~72%;六氯联苯去除率在50%以下。提高nZVI的添加浓度,高氯联苯(4~6 Cl)的去除率升高幅度最大。

RGO-nZVI作为强化剂,土壤中二氯联苯和三氯联苯去除率维持在88%以上;四氯联苯和五氯联苯去除率为68%~75%;一氯联苯和六氯联苯去除率在50%以下。RGO-nZVI作为强化剂,在根际袋和非根际袋添加RGO-nZVI,PCBs去除率差异不明显;提高RGO-nZVI的添加浓度,一氯联苯的去除率明显升高。

Biochar作为强化剂,土壤中二氯联苯和三氯联苯去除率维持在88%以上;四氯联苯和五氯联苯去除率为72%~77%;一氯联苯和六氯联苯去除率在55%以下。Biochar作为强化剂,与在根际袋添加强化剂相比,非根际袋添加强化剂使一氯联苯去除率明显升高,其他PCBs(2~6 Cl)去除率明显下降;与种植物相比,一氯联苯去除率明显升高,其他PCBs(2~6 Cl)去除率下降;提高Biochar的添加浓度没有提高PCBs的去除率。

Biochar-nZVI作为强化剂,土壤中低氯联苯(1~3 Cl)去除率维持在83%以上;四氯联苯和五氯联苯去除率为70%~76%;六氯联苯去除率在50%以下。Biochar-nZVI作为强化剂,在根际袋和非根际袋添加Biochar-nZVI,PCBs去除率差异不明显,六氯联苯去除率在非根际袋添加强化剂时有所升高;与种植物相比,不种植物时土壤PCBs去除率差异不明显,其中六氯联苯去除率有所升高;提高Biochar-nZVI的添加浓度没有降低土壤PCBs的去除率。

11.8 小结

本文分别制备和表征了nZVI、Biochar、GO、Biochar-nZVI和RGO-nZVI,并分别添加于土壤中进行强化植物修复Cd-PCBs复合污染土壤的盆栽试验。结果发现,高浓度nZVI、Biochar-nZ-

VI 和 RGO-nZVI 对孔雀草具有毒害作用，对孔雀草生长不利，说明 GO 和 Biochar 负载后并没有明显减弱纳米铁的不利影响。GO 作为强化剂时提高了植物生物量；低浓度 RGO-nZVI 作为强化剂时也提高了植物生物量；nZVI、Biochar 和 Biochar-nZVI 作为强化剂时没有明显提高植物生物量，反而抑制植物生长。相比其他强化剂，GO 作为强化剂时，土壤中 PCBs 的去除率较高，而且是一种植物友好型土壤修复强化剂。

参 考 文 献

[1] Fellet G, Marchiol L, Perosa D, et al. The application of phytoremediation technology in a soil contaminated by pyrite cinders [J]. Ecological Engineering, 2007, 31 (3): 207-214.
[2] Zhou Q, Song Y. Remediation of contaminated soils: principles and methods [M]. Beijing: Science Press, 2004.
[3] Dahmani-Muller H, Van Oort F, Gelie B, et al. Strategies of heavy metal uptake by three plant species growing near a metal smelter [J]. Environmental Pollution, 2000, 109 (2): 231-238.
[4] McGrath S P, Zhao F J. Phytoextraction of metals and metalloids from contaminated soils [J]. Current Opinion in Biotechnology, 2003, 14 (3): 277-282.
[5] Baker A, Brooks R. Terrestrial higher plants which hyperaccumulate metallic elements: a review of their distribution, ecology and phytochemistry [J]. Biorecovery, 1989, 1 (2): 81-126.
[6] Krämer U. Metal hyperaccumulation in plants [J]. Annual review of plant biology, 2010 (61): 517-534.
[7] Visoottiviseth P, Francesconi K, Sridokchan W. The potential of *Thai indigenous* plant species for the phytoremediation of arsenic contaminated land [J]. Environmental Pollution, 2002, 118 (3): 453-461.
[8] Hasegawa I, Terada E, Sunairi M, et al. Genetic improvement of heavy metal tolerance in plants by transfer of the yeast metallothionein gene (CUP1) [M]//Plant Nutrition for Sustainable Food Production and Environment. Tokyo: Springer, 1997: 391-395.
[9] 杜俊杰, 李娜, 吴永宁, 等. 蔬菜对重金属的积累差异及低积累蔬菜的研究进展[J]. 农业环境科学学报, 2019, 38 (6): 1193-1201.
[10] Brooks R, Lee J, Reeves R, et al. Detection of nickeliferous rocks by

analysis of herbarium specimens of indicator plants [J]. Journal of Geochemical Exploration, 1977, 7: 49-57.

[11] Salt D E, Smith R, Raskin I. Phytoremediation [J]. Annual review of plant biology, 1998, 49 (1): 643-668.

[12] Wei S, Zhou Q, Wang X, et al. A newly-discovered Cd-hyperaccumulator *Solatium nigrum* L [J]. Chinese Science Bulletin, 2005, 50 (1): 33-38.

[13] Karimi N, Ghaderian S M, Raab A, et al. An arsenic-accumulating, hypertolerant brassica, *Isatis capadocica* [J]. New Phytologist, 2009, 184 (1): 41-47.

[14] Macnair M R. The hyperaccumulation of metals by plants [J]. Advances in Botanical Research, 2003 (40): 63-105.

[15] 杜俊杰,周启星,李娜,等.超积累植物修复重金属污染土壤的研究进展[J].贵州农业科学, 2018, 46 (5): 64-72.

[16] Wei S, Zhou Q. Identification of weed species with hyperaccumulative characteristics of heavy metals [J]. Progress in Natural Science, 2004, 14 (6): 495-503.

[17] Baker A, McGrath S, Sidoli C, et al. The possibility of in situ heavy metal decontamination of polluted soils using crops of metal-accumulating plants [J]. Resources, Conservation and Recycling, 1994, 11 (1): 41-49.

[18] Rascio N, Navari-Izzo F. Heavy metal hyperaccumulating plants: how and why do they do it? And what makes them so interesting [J]. Plant Science, 2011, 180 (2): 169-181.

[19] Malaisse F, Gregoire J, Brooks R, et al. *Aeolanthus biformifolius* De Wild. : a hyperaccumulator of copper from Zaire [J]. Science, 1978, 199 (4331): 887-888.

[20] Reeves R, Brooks R. Hyperaccumulation of lead and zinc by two metallophytes from mining areas of Central Europe [J]. Environmental Pollution Series A, Ecological and Biological, 1983, 31 (4): 277-285.

[21] Chaney R L, Malik M, Li Y M, et al. Phytoremediation of soil metals [J]. Current Opinion in Biotechnology, 1997, 8 (3): 279-284.

参 考 文 献

[22] Salt D. Phytoextraction: present applications and future promise [C]. Environmental Science and Pollution Control Series, 2000: 729-744.

[23] Wei S, Zhou Q, Saha U K, et al. Identification of a Cd accumulator *Conyza canadensis* [J]. Journal of Hazardous Materials, 2009, 163 (1): 32-35.

[24] Sun Y, Zhou Q, Wang L, et al. Cadmium tolerance and accumulation characteristics of *Bidens pilosa* L. as a potential Cd-hyperaccumulator [J]. Journal of Hazardous Materials, 2009, 161 (2): 808-814.

[25] 安鑫龙, 周启星. 大型真菌对重金属的生物富集作用及生态修复[J]. 应用生态学报, 2007 (8): 1897-1902.

[26] 魏树和, 周启星, 张凯松, 等. 根际圈在污染土壤修复中的作用与机理分析[J]. 应用生态学报, 2003 (1): 143-147.

[27] 智杨, 孙挺, 周启星, 等. 铅低积累大豆的筛选及铅对其豆中矿物营养元素的影响[J]. 环境科学学报, 2015, 35 (6): 1939-1945.

[28] Brooks R R. Plants that hyperaccumulate heavy metals: their role in phytoremediation, microbiology, archaeology, mineral exploration and phytomining [M]. New York: CAB International, 1998.

[29] 周启星, 宋玉芳. 植物修复的技术内涵及展望[J]. 安全与环境学报, 2001, 1 (3): 48-53.

[30] Guo S, Huang C, Bian Y, et al. On absorption and accumulation of six heavy metal elements of weeds in Jinhua suburb-survey on content of six heavy metal elements in weeds and soil [J]. J Shanghai Jiaotong University (Agr Sci), 2002 (20): 1-8.

[31] Salt D E, Blaylock M, Kumar N P, et al. Phytoremediation: a novel strategy for the removal of toxic metals from the environment using plants [J]. Nature Biotechnology, 1995, 13 (5): 468-474.

[32] Raskin I, Ensley B D. Phytoremediation of toxic metals [M]. Mew York: John Wiley and Sons, 2000.

[33] Küpper H, Zhao F J, McGrath S P. Cellular compartmentation of zinc in leaves of the hyperaccumulator Thlaspi caerulescens [J]. Plant Physiology, 1999, 119 (1): 305-312.

[34] Kupper H, Lombi E, Zhao F J, et al. Cellular compartmentation of cad-

mium and zinc in relation to other elements in the hyperaccumulator Arabidopsis halleri [J]. Planta, 2000, 212 (1): 75-84.

[35] Ross S M. Toxic metals in soil-plant systems [M]. Chichester: John Wiley & Sons ltd, 1994.

[36] Meharg A. Integrated tolerance mechanisms: constitutive and adaptive plant responses to elevated metal concentrations in the environment [J]. Plant, Cell & Environment, 1994, 17 (9): 989-993.

[37] Küpper H, Lombi E, Zhao F J, et al. Cellular compartmentation of cadmium and zinc in relation to other elements in the hyperaccumulator *Arabidopsis halleri* [J]. Planta, 2000, 212 (1): 75-84.

[38] Krämer U, Cotter-Howells J D, Charnock J M, et al. Free histidine as a metal chelator in plants that accumulate nickel [J]. Nature, 1996, 379: 635-638.

[39] Murphy A, Zhou J, Goldsbrough P B, et al. Purification and immunological identification of metallothioneins 1 and 2 from *Arabidopsis thaliana* [J]. Plant Physiology, 1997, 113 (4): 1293-1301.

[40] Pickering I J, Prince R C, George M J, et al. Reduction and coordination of arsenic in Indian mustard [J]. Plant Physiology, 2000, 122 (4): 1171-1177.

[41] Stephan U W, Schmidke I, Stephan V W, et al. The nicotianamine molecule is made-to-measure for complexation of metal micronutrients in plants [J]. Biometals, 1996, 9 (1): 84-90.

[42] Higuchi K, Suzuki K, Nakanishi H, et al. Cloning of nicotianamine synthase genes, novel genes involved in the biosynthesis of phytosiderophores [J]. Plant Physiology, 1999, 119 (2): 471-480.

[43] Kusznierewicz B, Baczek-Kwinta R, Bartoszek A, et al. The dose-dependent influence of zinc and cadmium contamination of soil on their uptake and glucosinolate content in white cabbage (*Brassica oleracea* var. *capitata* f. *alba*) [J]. Environmental Toxicology and Chemistry, 2012, 31 (11): 2482-2489.

[44] Freeman J L, Persans M W, Nieman K, et al. Increased glutathione biosynthesis plays a role in nickel tolerance in *Thlaspi* nickel hyperaccumula-

参 考 文 献

tors [J]. The Plant Cell Online, 2004, 16 (8): 2176-2191.

[45] Freeman J L, Salt D E. The metal tolerance profile of Thlaspi goesingense is mimicked in *Arabidopsis thaliana* heterologously expressing serine acetyl-transferase [J]. Bmc Plant Biology, 2007, 7 (1): 63.

[46] Wei S, Ma L Q, Saha U, et al. Sulfate and glutathione enhanced arsenic accumulation by arsenic hyperaccumulator *Pteris vittata* L. [J]. Environmental Pollution, 2010, 158 (5): 1530-1535.

[47] Robinson N J, Tommey A M, Kuske C, et al. Plant metallothioneins [J]. Biochemical Journal, 1993, 295 (1): 1.

[48] Küpper H, Mijovilovich A, Meyer-Klaucke W, et al. Tissue-and age-dependent differences in the complexation of cadmium and zinc in the cadmium/zinc hyperaccumulator *Thlaspi caerulescens* (Ganges ecotype) revealed by X-ray absorption spectroscopy [J]. Plant Physiology, 2004, 134 (2): 748-757.

[49] Sarret G, Saumitou-Laprade P, Bert V, et al. Forms of zinc accumulated in the hyperaccumulator *Arabidopsis halleri* [J]. Plant Physiology, 2002, 130 (4): 1815-1826.

[50] Lee J, Reeves R D, Brooks R R, et al. Isolation and identification of a citrato-complex of nickel from nickel-accumulating plants [J]. Phytochemistry, 1977, 16 (10): 1503-1505.

[51] Von Wirén N, Klair S, Bansal S, et al. Nicotianamine chelates both FeI-II and FeII. Implications for metal transport in plants [J]. Plant Physiology, 1999, 119 (3): 1107-1114.

[52] Freeman J, Salt D. The metal tolerance profile of *Thlaspi goesingense* is mimicked in *Arabidopsis thaliana* heterologously expressing serine acetyltransferase [J]. BMC Plant Biology, 2007, 7 (1): 63.

[53] Benzarti S, Hamdi H, Mohri S, et al. Response of antioxidative enzymes and apoplastic bypass transport in *Thlaspi Caerulescens* and *Raphanus Sativus* to cadmium stress [J]. International Journal of Phytoremediation, 2010, 12 (8): 733-744.

[54] Cunningham S D, Berti W R. Remediation of contaminated soils with green plants: an overview [J]. In Vitro Cellular & Developmental Biolo-

gy-Plant, 1993, 29 (4): 207-212.

[55] Khoudi H, Maatar Y, Gouiaa S, et al. Transgenic tobacco plants expressing ectopically wheat H^+-pyrophosphatase (H^+-PPase) gene *TaVP1* show enhanced accumulation and tolerance to cadmium [J]. Journal of Plant Physiology, 2012, 169 (1): 98-103.

[56] Wenzel W W, Jockwer F. Accumulation of heavy metals in plants grown on mineralised soils of the *Austrian Alps* [J]. Environmental Pollution, 1999, 104 (1): 145-155.

[57] Reeves R, Baker A. Studies on metal uptake by plants from serpentine and non-serpentine populations of *Thlaspi goesingense* Halacsy (Cruciferae) [J]. New Phytologist, 1984, 98 (1): 191-204.

[58] Ma L Q, Komar K M, Tu C, et al. A fern that hyperaccumulates arsenic [J]. Nature, 2001, 409 (6820): 579-579.

[59] Chen T, Wei C, Huang Z, et al. Arsenic hyperaccumulator *Pteris vittata* L. and its arsenic accumulation [J]. Chinese Science Bulletin, 2002, 47 (11): 902-905.

[60] 杨肖娥,龙新宪,倪吾钟,等.东南景天(*Sedum alfredii* H):一种新的锌超积累植物[J].科学通报,2002,47(13):1003-1006.

[61] 陈同斌,韦朝阳,黄泽春,等.砷超富集植物蜈蚣草及其对砷的富集特征[J].科学通报,2002,47(3):207-210.

[62] 韦朝阳,陈同斌,黄泽春,等.大叶井口边草:一种新发现的富集砷的植物[J].生态学报,2002,22(5):777-778.

[63] 刘威,束文圣,蓝崇钰.宝山堇菜(*Viola baoshanensis*):一种新的镉超富集植物[J].科学通报,2003,48(19):2046-2049.

[64] 薛生国,陈英旭,林琦,等.中国首次发现的锰超积累植物:商陆[J].生态学报,2003,23(5):935-937.

[65] 束文圣,杨开颜,张志权,等.湖北铜绿山古铜矿冶炼渣植被与优势植物的重金属含量研究[J].应用与环境生物学报,2001(1):7-12.

[66] 张学洪,罗亚平,黄海涛,等.一种新发现的湿生铬超积累植物:李氏禾(*Leersia hexandra* Swartz)[J].生态学报,2006,26(3):950-953.

[67] 魏树和,周启星,王新,等.一种新发现的镉超积累植物龙葵(*Solanum nigrum* L)[J].科学通报,2004,49(24):2568-2573.

参 考 文 献

[68] Van der Ent A, Baker A J, Reeves R D, et al. Hyperaccumulators of metal and metalloid trace elements: facts and fiction [J]. Plant and Soil, 2013, 362 (1-2): 319-334.

[69] Baker A, Proctor J. The influence of cadmium, copper, lead, and zinc on the distribution and evolution of metallophytes in the British Isles [J]. Plant Systematics and Evolution, 1990, 173 (1-2): 91-108.

[70] 周启星, 宋玉芳. 污染土壤修复原理与方法 [M]. 北京: 科学出版社, 2004.

[71] McIntyre T. Phytoremediation of heavy metals from soils [M]. Berlin: Springer, 2003: 97-123.

[72] Wei S, Zhou Q, Zhan J, et al. Poultry manured *Bidens tripartite* L. extracting Cd from soil-potential for phytoremediating Cd contaminated soil [J]. Bioresource Technology, 2010, 101 (22): 8907-8910.

[73] Ensley B D, Blaylock M J, Dushenkov S, et al. Inducing hyperaccumulation of metals in plant shoots: US5917117A [P]. 1999.

[74] Marques A P, Oliveira R S, Samardjieva K A, et al. EDDS and EDTA-enhanced zinc accumulation by *solanum nigrum* inoculated with arbuscular mycorrhizal fungi grown in contaminated soil [J]. Chemosphere, 2008, 70 (6): 1002-1014.

[75] Quartacci M F, Irtelli B, Baker A J, et al. The use of NTA and EDDS for enhanced phytoextraction of metals from a multiply contaminated soil by *Brassica carinata* [J]. Chemosphere, 2007, 68 (10): 1920-1928.

[76] Evangelou M W H, Daghan H, Schaeffer A. The influence of humic acids on the phytoextraction of cadmium from soil [J]. Chemosphere, 2004, 57 (3): 207-213.

[77] Vamerali T, Bandiera M, Mosca G. Field crops for phytoremediation of metal-contaminated land: a review [J]. Environmental Chemistry Letters, 2010, 8 (1): 1-17.

[78] Halim M, Conte P, Piccolo A. Potential availability of heavy metals to phytoextraction from contaminated soils induced by exogenous humic substances [J]. Chemosphere, 2003, 52 (1): 265-275.

[79] Sadowsky M. Phytoremediation: past promises and future practices [R].

Proceedings of the 8th International Symposium on Microbial Ecology. Halifax: 1999.

[80] Baum C, Hrynkiewicz K, Leinweber P, et al. Heavy-metal mobilization and uptake by mycorrhizal and nonmycorrhizal willows (Salix× dasyclados) [J]. Journal of Plant Nutrition and Soil Science, 2006, 169 (4): 516-522.

[81] Whiting S N, Leake J R, McGrath S P, et al. Hyperaccumulation of Zn by Thlaspi caerulescens can ameliorate Zn toxicity in the rhizosphere of cocropped Thlaspi arvense [J]. Environmental science & technology, 2001, 35 (15): 3237-3241.

[82] Belimov A, Hontzeas N, Safronova V, et al. Cadmium-tolerant plant growth-promoting bacteria associated with the roots of Indian mustard (Brassica juncea L. Czern.) [J]. Soil Biology and Biochemistry, 2005, 37 (2): 241-250.

[83] Dimkpa C O, Svatoš A, Dabrowska P, et al. Involvement of siderophores in the reduction of metal-induced inhibition of auxin synthesis in Streptomyces spp [J]. Chemosphere, 2008, 74 (1): 19-25.

[84] Sheng X F, Xia J J. Improvement of rape (Brassica napus) plant growth and cadmium uptake by cadmium-resistant bacteria [J]. Chemosphere, 2006, 64 (6): 1036-1042.

[85] Hanikenne M, Nouet C. Metal hyperaccumulation and hypertolerance: a model for plant evolutionary genomics [J]. Current opinion in plant biology, 2011, 14 (3): 252-259.

[86] LeDuc D L, Terry N. Phytoremediation of toxic trace elements in soil and water [J]. Journal of Industrial Microbiology and Biotechnology, 2005, 32 (11-12): 514-520.

[87] De Souza M P, Pilon-Smits E A, Lytle C M, et al. Rate-limiting steps in selenium assimilation and volatilization by Indian mustard [J]. Plant Physiology, 1998, 117 (4): 1487-1494.

[88] Arazi T, Sunkar R, Kaplan B, et al. A tobacco plasma membrane calmodulin-binding transporter confers Ni^{2+} tolerance and Pb^{2+} hypersensitivity in transgenic plants [J]. The Plant Journal, 1999, 20 (2): 171-182.

参 考 文 献

[89] Hirschi K D, Korenkov V D, Wilganowski N L, et al. Expression of *Arabidopsis CAX2* in tobacco, altered metal accumulation and increased manganese tolerance [J]. Plant Physiology, 2000, 124 (1): 125-134.

[90] Evans K M, Gatehouse J A, Lindsay W P, et al. Expression of the pea metallothionein-like gene PsMT A in *Escherichia coli* and *Arabidopsis thaliana* and analysis of trace metal ion accumulation: implications for PsMT A function [J]. Plant Molecular Biology, 1992, 20 (6): 1019-1028.

[91] Rauser W E. Phytochelatins and related peptides. Structure, biosynthesis, and function [J]. Plant Physiology, 1995, 109 (4): 1141-1149.

[92] De la Fuente J M, Ramírez-Rodríguez V, Cabrera-Ponce J L, et al. Aluminum tolerance in transgenic plants by alteration of citrate synthesis [J]. Science, 1997, 276 (5318): 1566-1568.

[93] Goto F, Yoshihara T, Shigemoto N, et al. Iron fortification of rice seed by the soybean ferritin gene [J]. Nature Biotechnology, 1999, 17 (3): 282-286.

[94] Daniell H, Dhingra A. Multigene engineering: dawn of an exciting new era in biotechnology [J]. Current Opinion in Biotechnology, 2002, 13 (2): 136-141.

[95] Daniell H. Molecular strategies for gene containment in transgenic crops [J]. Nature Biotechnology, 2002, 20 (6): 581-586.

[96] Pilon-Smits E. Phytoremediation [J]. Annual Review of Plant Biology, 2005 (56): 15-39.

[97] Ma Y, Dickinson N M, Wong M H. Interactions between earthworms, trees, soil nutrition and metal mobility in amended Pb/Zn mine tailings from Guangdong, China [J]. Soil Biology and Biochemistry, 2003, 35 (10): 1369-1379.

[98] Ireland M P. The effect of the earthworm Dendrobaena rubida on the solubility of lead, zinc and calcium in heavy metal contaminated soil in Wales [J]. European Journal of Soil Science, 1975, 26 (3): 313-318.

[99] 陈建秀, 麻智春, 严海娟, 等. 跳虫在土壤生态系统中的作用[J]. 生物多样性, 2007 (2): 154-161.

[100] McGrath S, Lombi E, Gray C, et al. Field evaluation of Cd and Zn phytoextraction potential by the hyperaccumulators *Thlaspi caerulescens* and *Arabidopsis halleri* [J]. Environmental Pollution, 2006, 141 (1): 115-125.

[101] Wei S, Teixeira da Silva J A, Zhou Q. Agro-improving method of phytoextracting heavy metal contaminated soil [J]. Journal of Hazardous Materials, 2008, 150 (3): 662-668.

[102] Robinson B H, Leblanc M, Petit D, et al. The potential of *Thlaspi caerulescens* for phytoremediation of contaminated soils [J]. Plant and Soil, 1998, 203 (1): 47-56.

[103] Wu L, Luo Y, Xing X, et al. EDTA-enhanced phytoremediation of heavy metal contaminated soil with *Indian mustard* and associated potential leaching risk [J]. Agriculture, Ecosystems & Environment, 2004, 102 (3): 307-318.

[104] Wei S, Wang S, Zhou Q, et al. Potential of *Taraxacum mongolicum* Hand-Mazz for accelerating phytoextraction of cadmium in combination with eco-friendly amendments [J]. Journal of Hazardous Materials, 2010, 181 (1): 480-484.

[105] Sung M, Lee C Y, Lee S Z. Combined mild soil washing and compost-assisted phytoremediation in treatment of silt loams contaminated with copper, nickel, and chromium [J]. Journal of Hazardous Materials, 2011, 190 (1): 744-754.

[106] Li Y M, Chaney R, Brewer E, et al. Development of a technology for commercial phytoextraction of nickel: economic and technical considerations [J]. Plant and Soil, 2003, 249 (1): 107-115.

[107] Gupta S, Herren T, Wenger K, et al. In situ gentle remediation measures for heavy metal-pol luted soils [M]. Boca Raton: Lewis, 2010.

[108] Wei S, Zhou Q X. Phytoremediation of cadmium-contaminated soils by rorippa globosa using two-phase planting [J]. Environmental Science and Pollution Research, 2006, 13 (3): 151-155.

[109] Ji P, Sun T, Song Y, et al. Strategies for enhancing the phytoremediation of cadmium-contaminated agricultural soils by *Solanum nigrum* L.

[J]. Environmental Pollution, 2011, 159 (3): 762-768.

[110] 杨元根. 稀土元素在红壤中的环境效应研究[J]. 土壤通报, 1998, 29 (3): 129-132.

[111] 何振立. 污染及有益元素的土壤化学平衡[M]. 北京: 中国环境科学出版社, 1998.

[112] Dahmani-Muller H, Van Oort F, Balabane M. Metal extraction by *Arabidopsis halleri* grown on an unpolluted soil amended with various metal-bearing solids: a pot experiment [J]. Environmental Pollution, 2001, 114 (1): 77-84.

[113] Reeves R D, Baker A J. Metal-accumulating plants [M]. New York: John Wiley and Sons, 2000.

[114] Shirong T. Distributional characteristics of hyperaccumulators at time and space as well as in family and genus [J]. Journal of Ecology and Rural Environment, 2001, 17 (4): 56-60.

[115] Gans J, Wolinsky M, Dunbar J. Computational improvements reveal great bacterial diversity and high metal toxicity in soil [J]. Science, 2005, 309 (5739): 1387-1390.

[116] Spalt E W, Kissel J C, Shirai J H, et al. Dermal absorption of environmental contaminants from soil and sediment: a critical review [J]. Journal of Exposure Science and Environmental Epidemiology, 2009, 19 (2): 119-148.

[117] 孙铁珩, 李培军, 周启星. 土壤污染形成机理与修复技术[M]. 北京: 科学出版社, 2005.

[118] 李冬梅. 西安市蔬菜基地持久性有机污染物（POPs）残留状况研究[D]. 西安: 陕西师范大学, 2008.

[119] 蔡全英, 莫测辉. 土壤中多氯代二噁英（PCDDs）的研究进展[J]. 农村生态环境, 1999, 15 (2): 41-45.

[120] 柳絮, 范仲学, 张斌, 等. 我国土壤镉污染及其修复研究[J]. 山东农业科学, 2008 (6): 94-97.

[121] 罗绪强, 王世杰, 张桂玲. 土壤镉污染及其生物修复研究进展[J]. 山地农业生物学报, 2008, 27 (4): 357-361.

[122] 周启星, 宋玉芳. 污染土壤修复原理与方法[M]. 北京: 科学出版

社，2004.
[123] 陈同斌，韦朝阳，黄泽春，等．砷超富集植物蜈蚣草及其对砷的富集特征[J]．科学通报，2002（3）：207-210.
[124] 杨肖娥，傅承新．东南景天（Sedum alfreii H）：一种新的锌超积累植物[J]．科学通报，2002，47（13）：1003-1006.
[125] Gan S，Lau E V，Ng H K. Remediation of soils contaminated with polycyclic aromatic hydrocarbons（PAHs）[J]. Journal of Hazardous Materials，2009，172（2-3）：532-549.
[126] Li N，Du J J，Wu D，et al. Recent advances in facile synthesis and applications of covalent organic framework materials as superior adsorbents in sample pretreatment [J]. Trac-Trends in Analytical Chemistry，2018（108）：154-166.
[127] Casey W. The fate of chlorine in soils [J]. Science，2002，295（5557）：985-986.
[128] 李森，陈家军，孟占利．多氯联苯处理处置方法国内外研究进展[J]．中国环保产业，2004（2）：26-29.
[129] 薛荔栋，郎印海，刘爱霞，等．零价铁脱氯还原多氯联苯的研究进展[J]．环境科学与技术，2009，32（3）：78-82.
[130] Du J J，Zhou Q X. Preliminary study on effects of nanoscale amendments on hyperaccumulator Indian Marigold grown on co-contaminated soils [J]. Advanced Materials Research，2014（955-959）：243-247.
[131] Bixio D，Thoeye C，De Koning J，et al. Wastewater reuse in Europe [J]. Desalination，2006，187（1）：89-101.
[132] 甘晓初．低温条件下高品质石墨烯的合成及其在水污染处理中的应用[D]．合肥：中国科学技术大学，2011.
[133] 肖蓝，王祎龙，于水利，等．石墨烯及其复合材料在水处理中的应用[J]．化学进展，2013，25（2）：419-430.
[134] Novoselov K S，Fal V，Colombo L，et al. A roadmap for graphene [J]. Nature，2012，490（7419）：192-200.
[135] Wang J，Chen Z，Chen B. Adsorption of polycyclic aromatic hydrocarbons by graphene and graphene oxide nanosheets [J]. Environmental Science & Technology，2014，48（9）：4817-4825.

[136] Sun H, Liu S, Zhou G, et al. Reduced graphene oxide for catalytic oxidation of aqueous organic pollutants [J]. ACS Applied Materials & Interfaces, 2012, 4 (10): 5466-5471.

[137] Wang S, Sun H, Ang H-M, et al. Adsorptive remediation of environmental pollutants using novel graphene-based nanomaterials [J]. Chemical Engineering Journal, 2013 (226): 336-347.

[138] Du J, Hu X, Mu L, et al. Root exudates as natural ligands that alter the properties of graphene oxide and environmental implications thereof [J]. RSC Advances, 2015, 5 (23): 17615-17622.

[139] Du J, Hu X, Zhou Q. Graphene oxide regulates the bacterial community and exhibits property changes in soil [J]. RSC Advances, 2015, 5 (34): 27009-27017.

[140] 曲晨，刘伟，荣海钦，等. 纳米银的生物学特性及其潜在毒性的研究进展[J]. 环境与健康杂志, 2010, 27 (9): 842-845.

[141] Siddiqui M H, Al-Whaibi M H. Role of nano-SiO_2 in germination of tomato (*Lycopersicum esculentum* seeds Mill.) [J]. Saudi Journal of Biological Sciences, 2014, 21 (1): 13-17.

[142] Lin D, Xing B. Phytotoxicity of nanoparticles: inhibition of seed germination and root growth [J]. Environmental Pollution, 2007, 150 (2): 243-250.

[143] Park S, Ahn Y J. Multi-walled carbon nanotubes and silver nanoparticles differentially affect seed germination, chlorophyll content, and hydrogen peroxide accumulation in carrot (*Daucus carota* L.) [J]. Biocatalysis and Agricultural Biotechnology, 2016 (8): 257-262.

[144] Novoselov K S, Fal'ko V I, Colombo L, et al. A roadmap for graphene [J]. Nature, 2012, 490 (7419): 192-200.

[145] Li H, Song Z N, Zhang X J, et al. Ultrathin, molecular-sieving graphene oxide membranes for selective hydrogen separation [J]. Science, 2013, 342 (6154): 95-98.

[146] Hu X G, Zhou Q X. Health and Ecosystem Risks of Graphene [J]. Chemical Reviews, 2013, 113 (5): 3815-3835.

[147] Hu X, Zhou Q. Health and ecosystem risks of graphene [J]. Chemical

Reviews, 2013, 113 (5): 3815-3835.

[148] Tong Z H, Bischoff M, Nies L, et al. Impact of fullerene (C_{60}) on a soil microbial community [J]. Environ Sci Technol, 2007, 41 (8): 2985-2991.

[149] Kirkham M B. Cadmium in plants on polluted soils: effects of soil factors, hyperaccumulation, and amendments [J]. Geoderma, 2006, 137 (1-2): 19-32.

[150] Sangsuwan P, Prapagdee B. Cadmium phytoremediation performance of two species of *Chlorophytum* and enhancing their potentials by cadmium-resistant bacteria [J]. Environmental Technology & Innovation, 2021 (21): 101311.

[151] Ding C F, Zhang T L, Wang X X, et al. Effects of soil type and genotype on cadmium accumulation by rootstalk crops: implications for phytomanagement [J]. International Journal of Phytoremediation, 2014, 16 (10): 1018-1030.

[152] Niu H, Leng Y, Li X, et al. Behaviors of cadmium in rhizosphere soils and its interaction with microbiome communities in phytoremediation [J]. Chemosphere, 2021 (269): 128765.

[153] Wenhao Y, Li P, Rensing C, et al. Changes in metal availability and improvements in microbial properties after phytoextraction of a Cd, Zn and Pb contaminated soil [J]. Bulletin of Environmental Contamination and Toxicology, 2018 (101): 624-630.

[154] Epelde L, Becerril J M, Mijangos I, et al. Evaluation of the efficiency of a phytostabilization process with biological indicators of soil health [J]. Journal of Environmental Quality, 2009, 38 (5): 2041-2049.

[155] McGrath S P, Zhao F J. Phytoextraction of metals and metalloids from contaminated soils [J]. Current Opinion in Biotechnology, 2003, 14 (3): 277-282.

[156] Chen C, Wang X, Wang J. Phytoremediation of cadmium-contaminated soil by *Sorghum bicolor* and the variation of microbial community [J]. Chemosphere, 2019 (235): 985-994.

[157] Kramer U. Metal hyperaccumulation in plants [J] Annual Review of

Plant Biology, 2010, 61: 517-534.

[158] Lian J, Zhao L, Wu J, et al. Foliar spray of TiO_2 nanoparticles prevails over root application in reducing Cd accumulation and mitigating Cd-induced phytotoxicity in maize (Zea mays L.) [J]. Chemosphere, 2020 (239): 124794.

[159] Raza A, Habib M, Kakavand S N, et al. Phytoremediation of cadmium: physiological, biochemical, and molecular mechanisms [J]. Biology, 2020, 9 (7): 177.

[160] Dai H, Wei S, Skuza L, et al. Phytoremediation of two ecotypes cadmium hyperaccumulator *Bidens pilosa* L. sourced from clean soils [J]. Chemosphere, 2021 (273): 129652.

[161] Ramana S, Tripathi A K, Kumar A, et al. Evaluation of *Furcraea foetida* (L.) Haw. for phytoremediation of cadmium contaminated soils [J]. Environmental Science and Pollution Research, 2021, 28 (11): 14177-14181.

[162] Sun Y B, Zhou Q X, Xu Y M, et al. Phytoremediation for co-contaminated soils of benzo a pyrene (B [a] P) and heavy metals using ornamental plant *Tagetes patula* [J]. Journal of Hazardous Materials, 2011, 186 (2-3): 2075-2082.

[163] Liu H Y, Guo S S, Jiao K, et al. Bioremediation of soils co-contaminated with heavy metals and 2, 4, 5-trichlorophenol by fruiting body of *Clitocybe maxima* [J]. Journal of Hazardous Materials, 2015 (294): 121-127.

[164] Zehra A, Sahito Z A, Tong W, et al. Identification of high cadmium-accumulating oilseed sunflower (*Helianthus annuus*) cultivars for phytoremediation of an Oxisol and an Inceptisol [J]. Ecotoxicology and Environmental Safety, 2020 (187): 109857.

[165] Thijs S, Sillen W, Rineau F, et al. Towards an enhanced understanding of plant-microbiome interactions to improve phytoremediation: engineering the metaorganism [J]. Front Microbiol, 2016 (7) 341-341.

[166] Louca S, Parfrey L W, Doebeli M. Decoupling function and taxonomy in the global ocean microbiome [J]. Science, 2016, 353 (6305):

1272-1277.
[167] Gao Y, Zhou P, Mao L, et al. Effects of plant species coexistence on soil enzyme activities and soil microbial community structure under Cd and Pb combined pollution [J]. Journal of Environmental Sciences, 2010, 22 (7): 1040-1048.
[168] Hu L, Wang R, Liu X, et al. Cadmium phytoextraction potential of king grass (*Pennisetum sinese* Roxb.) and responses of rhizosphere bacterial communities to a cadmium pollution gradient [J]. Environmental Science and Pollution Research, 2018, 25 (22): 21671-21681.
[169] Hou F, Du J. Effects of soil properties on phytoextraction of Cd and the associated soil bacterial communities across four soil types [J]. Soil and Sediment Contamination: An International Journal, 2022, 31 (3): 282-292.
[170] Lu R K. Analytical methods of soil and agricultural chemistry [M]. Beijing: China Agricultural Science and Technology Press, 1999.
[171] Jiang R, Wang M, Chen W, et al. Ecological risk evaluation of combined pollution of herbicide siduron and heavy metals in soils [J]. Science of the Total Environment, 2018 (626): 1047-1056.
[172] Du J J, Hu X G, Zhou Q X. Graphene oxide regulates the bacterial community and exhibits property changes in soil [J]. RSC Advances, 2015, 5 (34): 27009-27017.
[173] Du J, Zhou Q, Wu J, et al. Vegetation alleviate the negative effects of graphene oxide on benzo [a] pyrene dissipation and the associated soil bacterial community [J]. Chemosphere, 2020 (253): 126725.
[174] Zeng F R, Ali S, Zhang H T, et al. The influence of pH and organic matter content in paddy soil on heavy metal availability and their uptake by rice plants [J]. Environmental Pollution, 2011, 159 (1): 84-91.
[175] Yang Z, Lu W, Long Y, et al. Assessment of heavy metals contamination in urban topsoil from Changchun City, China [J]. Journal of Geochemical Exploration, 2011, 108 (1): 27-38.
[176] Guo J, Yang J, Yang J, et al. Water-soluble chitosan enhances phytoremediation efficiency of cadmium by *Hylotelephium spectabile* in con-

taminated soils [J]. Carbohydrate Polymers, 2020 (246): 116559.
[177] Ashraf M A, Maah M J, Yusoff I. Chemical speciation and potential mobility of heavy metals in the soil of former tin mining catchment [J]. Scientific World Journal, 2012 (125608): 1-11.
[178] Fu J T, Yu D M, Chen X, et al. Recent research progress in geochemical properties and restoration of heavy metals in contaminated soil by phytoremediation [J]. Journal of Mountain Science, 2019 (16): 1-17.
[179] Karna R R, Noerpel M, Betts A R, et al. Lead and arsenic bioaccessibility and speciation as a function of soil particle size [J]. Journal of Environmental Quality, 2017, 46 (6): 1225-1235.
[180] Ferraz M C M A, Lourençlo J C N. The influence of organic matter content of contaminated soils on the leaching rate of heavy metals [J]. Environmental Progress, 2000, 19 (1): 53-58.
[181] Fang Y Y, Cao X D, Zhao L. Effects of phosphorus amendments and plant growth on the mobility of Pb, Cu, and Zn in a multi-metal-contaminated soil [J]. Environmental Science and Pollution Research, 2012, 19 (5): 1659-1667.
[182] Henriques I, Araújo S, Pereira A, et al. Combined effect of temperature and copper pollution on soil bacterial community: Climate change and regional variation aspects [J]. Ecotoxicology and Environmental Safety, 2015 (111): 153-159.
[183] Lauber C L, Strickland M S, Bradford M A, et al. The influence of soil properties on the structure of bacterial and fungal communities across land-use types [J]. Soil Biology and Biochemistry, 2008, 40 (9): 2407-2415.
[184] Shen C, Ge Y, Yang T, et al. Verrucomicrobial elevational distribution was strongly influenced by soil pH and carbon/nitrogen ratio [J]. Journal of Soils and Sediments, 2017, 17 (10): 2449-2456.
[185] Hoshino Y T, Morimoto S, Hayatsu M, et al. Effect of soil type and fertilizer management on archaeal community in upland field soils [J]. Microbes and Environments, 2011, 26 (4): 307-316.
[186] Chang H, Sun Z, Saito M, et al. Regulating infrared photoresponses in

reduced graphene oxide phototransistors by defect and atomic structure dontrol [J]. ACS Nano, 2013, 7 (7): 6310-6320.

[187] Kjellerup B V, Naff C, Edwards S J, et al. Effects of activated carbon on reductive dechlorination of PCBs by organohalide respiring bacteria indigenous to sediments [J]. Water Research, 2014 (52): 1-10.

[188] Ren G D, Teng Y, Ren W J, et al. Pyrene dissipation potential varies with soil type and associated bacterial community changes [J]. Soil Biology and Biochemistry, 2016 (103): 71-85.

[189] Chekol T, Vough L R, Chaney R L. Phytoremediation of polychlorinated biphenyl-contaminated soils: the rhizosphere effect [J]. Environment International, 2004, 30 (6): 799-804.

[190] Borja J, Taleon D M, Auresenia J, et al. Polychlorinated biphenyls and their biodegradation [J]. Process Biochemistry, 2005, 40 (6): 1999-2013.

[191] Correa P A, Lin L, Just C L, et al. The effects of individual PCB congeners on the soil bacterial community structure and the abundance of biphenyl dioxygenase genes [J]. Environment International, 2010, 36 (8): 901-906.

[192] Ding N, Guo H C, Hayat T, et al. Microbial community structure changes during Aroclor 1242 degradation in the rhizosphere of ryegrass (*Lolium multifiorum* L.) [J]. Fems Microbiology Ecology, 2009, 70 (2): 305-314.

[193] Hornbuckle K, Robertson L. Polychlorinated biphenyls (PCBs): sources, exposures, toxicities [J]. Environmental Science & Technology, 2010, 44 (8): 2749-2751.

[194] Van Aken B, Correa P A, Schnoor J L. Phytoremediation of polychlorinated biphenyls: new trends and promises [J]. Environmental Science & Technology, 2010, 44 (8): 2767-2776.

[195] Uhlik O, Jecna K, Mackova M, et al. Biphenyl-metabolizing bacteria in the rhizosphere of horseradish and bulk soil contaminated by polychlorinated biphenyls as revealed by stable isotope probing [J]. Applied and Environmental Microbiology, 2009, 75 (20): 6471-6477.

[196] Low J E, Aslund M L W, Rutter A, et al. Effect of plant age on PCB

accumulation by *Cucurbita pepo* ssp pepo [J]. Journal of Environmental Quality, 2010, 39 (1): 245-250.

[197] Petrić I, Bru D, Udiković-Kolić N, et al. Evidence for shifts in the structure and abundance of the microbial community in a long-term PCB-contaminated soil under bioremediation [J]. Journal of Hazardous Materials, 2011 (195): 254-260.

[198] Haritash A K, Kaushik C P. Biodegradation aspects of polycyclic aromatic hydrocarbons (PAHs): a review [J]. Journal of Hazardous Materials, 2009, 169 (1-3): 1-15.

[199] Liang Y, Meggo R, Hu D F, et al. Microbial community analysis of switchgrass planted and unplanted soil microcosms displaying PCB dechlorination [J]. Applied Microbiology and Biotechnology, 2015, 99 (15): 6515-6526.

[200] Tillmann S, Strompl C, Timmis K N, et al. Stable isotope probing reveals the dominant role of Burkholderia species in aerobic degradation of PCBs [J]. Fems Microbiology Ecology, 2005, 52 (2): 207-217.

[201] Komancova M, Jurcova I, Kochankova L, et al. Metabolic pathways of polychlorinated biphenyls degradation by *Pseudomonas* sp 2 [J]. Chemosphere, 2003, 50 (4): 537-543.

[202] Leigh M B, Prouzova P, Mackova M, et al. Polychlorinated biphenyl (PCB) -degrading bacteria associated with trees in a PCB-contaminated site [J]. Applied and Environmental Microbiology, 2006, 72 (4): 2331-2342.

[203] Seah S Y K, Ke J Y, Denis G, et al. Characterization of a C-C bond hydrolase from *Sphingomonas wittichii* RW1 with novel specificities towards polychlorinated biphenyl metabolites [J]. Journal of Bacteriology, 2007, 189 (11): 4038-4045.

[204] Field J A, Sierra-Alvarez R. Microbial transformation and degradation of polychlorinated biphenyls [J]. Environmental Pollution, 2008, 155 (1): 1-12.

[205] Liu J, He X X, Lin X R, et al. Ecological effects of combined pollution associated with e-waste recycling on the composition and diversity of soil

microbial communities [J]. Environmental Science & Technology, 2015, 49 (11): 6438-6447.

[206] Lemanowicz J, Haddad S A, Bartkowiak A, et al. The role of an urban park's tree stand in shaping the enzymatic activity, glomalin content and physicochemical properties of soil [J]. Science of the Total Environment, 2020 (741): 140446.

[207] Janvier C, Villeneuve F, Alabouvette C, et al. Soil health through soil disease suppression: which strategy from descriptors to indicators [J]. Soil Biology and Biochemistry, 2007, 39 (1): 1-23.

[208] Ti Q, Gu C, Liu C, et al. Comparative evaluation of influence of aging, soil properties and structural characteristics on bioaccessibility of polychlorinated biphenyls in soil [J]. Chemosphere, 2018 (210): 941-948.

[209] Lehtinen T, Mikkonen A, Sigfusson B, et al. Bioremediation trial on aged PCB-polluted soils: a bench study in Iceland [J]. Environmental Science and Pollution Research, 2014, 21 (3): 1759-1768.

[210] Starr J M, Li W, Graham S E, et al. Is food type important for in vitro post ingestion bioaccessibility models of polychlorinated biphenyls sorbed to soil [J]. Science of the Total Environment, 2020 (704): 135421.

[211] Girvan M S, Bullimore J, Pretty J N, et al. Soil type is the primary determinant of the composition of the total and active bacterial communities in arable soils [J]. Applied and Environmental Microbiology, 2003, 69 (3): 1800-1809.

[212] Xu Y X, Wang G H, Jin J, et al. Bacterial communities in soybean rhizosphere in response to soil type, soybean genotype, and their growth stage [J]. Soil Biology and Biochemistry, 2009, 41 (5): 919-925.

[213] Du J, Hou F, Zhou Q. Response of soil enzyme activity and soil bacterial community to PCB dissipation across different soils [J]. Chemosphere, 2021 (283): 131229.

[214] Trasar-Cepeda C, Camiña F, Leirós M C, et al. An improved method to measure catalase activity in soils [J]. Soil Biology and Biochemistry, 1999, 31 (3): 483-485.

[215] Arif M S, Riaz M, Shahzad S M, et al. Fresh and composted industrial

sludge restore soil functions in surface soil of degraded agricultural land [J]. Science of the Total Environment, 2018 (619-620): 517-527.

[216] Floch C, Alarcon-Gutiérrez E, Criquet S. ABTS assay of phenol oxidase activity in soil [J]. Journal of Microbiological Methods, 2007, 71 (3): 319-324.

[217] Gao M, Dong Y, Zhang Z, et al. Effect of dibutyl phthalate on microbial function diversity and enzyme activity in wheat rhizosphere and non-rhizosphere soils [J]. Environmental Pollution, 2020 (265): 114800.

[218] Du J, Zhou Q, Wu J, et al. Soil bacterial communities respond differently to graphene oxide and reduced graphene oxide after 90 days of exposure [J]. Soil Ecology Letters, 2020, 2 (3): 176-179.

[219] MEPC. Standard of soil quality assessment for exhibition sites (HJ 350-2007) [M]. Beijing: Ministry of Environmental Protection of the People's Republic of China, 2007.

[220] Chen Z Y, Zhang W, Wang G, et al. Bioavailability of soil-sorbed tetracycline to *Escherichia coli* under unsaturated conditions [J]. Environmental Science & Technology, 2017, 51 (11): 6165-6173.

[221] Guimarães A C D, Mendes K F, Reis F C D, et al. Role of soil physicochemical properties in quantifying the fate of diuron, hexazinone, and metribuzin [J]. Environmental Science & Pollution Research, 2018, 25 (5): 1-15.

[222] Zhang A P, Chen Z Y, Ahrens L, et al. Concentrations of DDTs and enantiomeric fractions of chiral DDTs in agricultural soils from Zhejiang Province, China, and correlations with total organic carbon and pH [J]. Journal of Agricultural and Food Chemistry, 2012, 60 (34): 8294-8301.

[223] An C J, Huang G H, Yu H, et al. Effect of short-chain organic acids and pH on the behaviors of pyrene in soil-water system [J]. Chemosphere, 2010, 81 (11): 1423-1429.

[224] Chien S W C, Chen S H, Li C J. Effect of soil pH and organic matter on the adsorption and desorption of pentachlorophenol [J]. Environmental Science and Pollution Research, 2018, 25 (6): 5269-5279.

[225] Paz-Ferreiro J, Trasar-Cepeda C, Leirós M C, et al. Biochemical properties in managed grassland soils in a temperate humid zone: modifications of soil quality as a consequence of intensive grassland use [J]. Biology and Fertility of Soils, 2009, 45 (7): 711-722.

[226] Lee S H, Xiong J Q, Ru S, et al. Toxicity of benzophenone-3 and its biodegradation in a freshwater microalga *Scenedesmus obliquus* [J]. Journal of Hazardous Materials, 2020 (389): 122149.

[227] Gao P T, Guo L, Sun J, et al. Accelerating waste sludge hydrolysis with alkyl polyglucose pretreatment coupled with biological process of thermophilic bacteria: hydrolytic enzyme activity and organic matters transformation [J]. J Environ Manage, 2019 (247): 161-168.

[228] Shen H, Li W, Graham S E, et al. The role of soil and house dust physicochemical properties in determining the post ingestion bioaccessibility of sorbed polychlorinated biphenyls [J]. Chemosphere, 2019 (217): 1-8.

[229] Lu L, Xing D, Ren N. Pyrosequencing reveals highly diverse microbial communities in microbial electrolysis cells involved in enhanced H_2 production from waste activated sludge [J]. Water Research, 2012, 46 (7): 2425-2434.

[230] Hiraishi A. Biodiversity of dehalorespiring bacteria with special emphasis on polychlorinated biphenyl/dioxin dechlorinators [J]. Microbes and Environments, 2008, 23 (1): 1-12.

[231] Chang Y C, Takada K, Choi D, et al. Isolation of biphenyl and polychlorinated biphenyl-degrading bacteria and their degradation pathway [J]. Applied Biochemistry and Biotechnology, 2013, 170 (2): 381-398.

[232] Papale M, Giannarelli S, Francesconi S, et al. Enrichment, isolation and biodegradation potential of psychrotolerant polychlorinated-biphenyl degrading bacteria from the Kongsfjorden (Svalbard Islands, High Arctic Norway) [J]. Marine Pollution Bulletin, 2017, 114 (2): 849-859.

[233] Xu Y, Teng Y, Wang X, et al. Exploring bacterial community structure and function associated with polychlorinated biphenyl biodegradation

in two hydrogen-amended soils [J]. Science of the Total Environment, 2020 (745): 140839.

[234] Chen Q L, Ding J, Zhu D, et al. Rare microbial taxa as the major drivers of ecosystem multifunctionality in long-term fertilized soils [J]. Soil Biology and Biochemistry, 2020 (141): 107686.

[235] Chen Y M, Zhan T L, Wei H Y, et al. Aging and compressibility of municipal solid wastes [J]. Waste Manage, 2009, 29 (1): 86-95.

[236] Zhao Y, Song L, Huang R, et al. Recycling of aged refuse from a closed landfill [J]. Waste Management & Research, 2007, 25 (2): 130-138.

[237] Liu Q M, Li Q B, Wang N, et al. Bioremediation of petroleum-contaminated soil using aged refuse from landfills [J]. Waste Manage, 2018 (77): 576-585.

[238] Li G K, Liu X F, Han M, et al. Selecting tolerant grass seedlings and analyzing the possibility for using aged refuse as sward soil [J]. Ecotoxicology and Environmental Safety, 2010, 73 (4): 620-625.

[239] Li D M, Zhang X J, Li G H, et al. Effects of heavy metal ions on germination and physiological activity of festuca arundinacea seed [J]. Pratacultural Science, 2008 (25): 98-102.

[240] Sun Y B, Zhou Q X, Xu Y M, et al. Phytoremediation for co-contaminated soils of benzo [a] pyrene (B [a] P) and heavy metals using ornamental plant Tagetes patula [J]. Journal of Hazardous Materials, 2011, 186 (2-3): 2075-2082.

[241] Wei J L, Lai H Y, Chen Z S. Chelator effects on bioconcentration and translocation of cadmium by hyperaccumulators, *Tagetes patula* and *Impatiens walleriana* [J]. Ecotoxicology and Environmental Safety, 2012 (84): 173-178.

[242] Goswami S, Das S. Screening of cadmium and copper phytoremediation ability of *Tagetes erecta*, using biochemical parameters and SEM-EDX analysis [J]. Environmental Toxicology & Chemistry, 2017, 36 (9): 2533-2542.

[243] Sathya V, Mahimairaja S, Bharani A, et al. Influence of soil bioamend-

ments on the availabilty of nickel and phytoextraction capability of marigold from the contaminated soil [J]. International Journal of Plant & Soil Science, 2020, 31 (5): 1-12.

[244] Madanan M T, Shah I K, Varghese G K, et al. Application of Aztec Marigold (*Tagetes erecta* L.) for phytoremediation of heavy metal polluted lateritic soil [J]. Environmental Chemistry and Ecotoxicology, 2021 (3): 17-22.

[245] Chen Y, Jiang Y, Huang H, et al. Long-term and high-concentration heavy-metal contamination strongly influences the microbiome and functional genes in Yellow River sediments [J]. Science of the Total Environment, 2018 (637-638): 1400-1412.

[246] Guarino F, Improta G, Triassi M, et al. Effects of zinc pollution and compost amendment on the root microbiome of a metal tolerant poplar clone [J]. Front Microbiol, 2020 (11): 1677.

[247] Huang Y, Xu D, Huang L Y, et al. Responses of soil microbiome to steel corrosion [J]. NPJ Biofilms Microbiomes, 2021, 7 (1): 6.

[248] Jiao S, Chen W, Wei G. Resilience and assemblage of soil microbiome in response to chemical contamination combined with plant growth [J]. Applied and Environmental Microbiology, 2019, 85 (6): e02523-18.

[249] Gałązka A, Grzadziel J, Gałązka R, et al. Genetic and functional diversity of bacterial microbiome in soils with long term impacts of petroleum hydrocarbons [J]. Front Microbiol, 2018 (9): 1923.

[250] Parajuli A, Gronroos M, Kauppi S, et al. The abundance of health-associated bacteria is altered in PAH polluted soils-Implications for health in urban areas [J]. PLoS One, 2017, 12 (11): e0187852.

[251] Hou F, Du J J, Yuan Y, et al. Analysis of microbial communities in aged refuse based on 16S sequencing [J]. Sustainability, 2021, 13 (8): 4111.

[252] Pandey N, Bhatt R. Role of soil associated *Exiguobacterium* in reducing arsenic toxicity and promoting plant growth in Vigna radiata [J]. Eur J Soil Biol, 2016 (75): 142-150.

[253] Li G, Hou F, Guo Z, et al. Analyzing nutrient distribution in different

particle-size municipal aged refuse [J]. Waste Manage, 2011, 31 (11): 2203-2207.

[254] Mocé-Llivina L, Muniesa M, Pimenta-Vale H, et al. Survival of bacterial indicator species and bacteriophages after thermal treatment of sludge and sewage [J]. Applied and Environmental Microbiology, 2003, 69 (3): 1452-1456.

[255] Drazkiewicz M, Baszynski T. Interference of nickel with the photosynthetic apparatus of *Zea mays* [J]. Ecotoxicology and Environmental Safety, 2010, 73 (5): 982-986.

[256] Bhaduri A M, Fulekar M H. Antioxidant enzyme responses of plants to heavy metal stress [J]. Reviews in Environmental Science and Bio-Technology, 2012, 11 (1): 55-69.

[257] Goswami S, Das S. Copper phytoremediation potential of *Calandula officinalis* L. and the role of antioxidant enzymes in metal tolerance [J]. Ecotoxicology and Environmental Safety, 2016 (126): 211-218.

[258] Hristozkova M, Geneva M, Stancheva I, et al. Contribution of arbuscular mycorrhizal fungi in attenuation of heavy metal impact on *Calendula officinalis* development [J]. Applied Soil Ecology, 2016 (101): 57-63.

[259] Lajayer B A, Ghorbanpour M, Nikabadi S. Heavy metals in contaminated environment: destiny of secondary metabolite biosynthesis, oxidative status and phytoextraction in medicinal plants [J]. Ecotoxicology and Environmental Safety, 2017, 145: 377-390.

[260] Apel K, Hirt H. Reactive oxygen species: metabolism, oxidative stress, and signal transduction [J]. Annual Review of Plant Biology, 2004, 55 (1): 373-399.

[261] Sharma P, Jha A B, Dubey R S, et al. Reactive oxygen species, oxidative damage, and antioxidative defense mechanism in plants under stressful conditions [J]. Journal of Botany, 2012 (2012): 1-26.

[262] Abbas G, Murtaza B, Bibi I, et al. Arsenic uptake, toxicity, detoxification, and speciation in plants: physiological, biochemical, and molecular aspects [J]. International Journal of Environmental Research and Public Health, 2018, 15 (1): 29301332.

[263] Zhang X Y, Zhang H Q, Lou X, et al. Mycorrhizal and non-mycorrhizal *Medicago truncatula* roots exhibit differentially regulated NADPH oxidase and antioxidant response under Pb stress [J]. Environmental and Experimental Botany, 2019 (164): 10-19.

[264] Ding B, Shi G, Xu Y, et al. Physiological responses of *Alternanthera philoxeroides* (Mart.) Griseb leaves to cadmium stress [J]. Environmental Pollution, 2007, 147 (3): 800-803.

[265] Yadav B, Johri A K, Dua M. Metagenomic analysis of the microbial diversity in solid waste from Okhla Landfill, New Delhi, India [J]. Microbiology Resource Announcements, 2020, 9 (46): e00921-20.

[266] Sekhohola Dlamini L, Selvarajan R, Ogola H J O, et al. Community diversity metrics, interactions, and metabolic functions of bacteria associated with municipal solid waste landfills at different maturation stages [J]. Microbiology Open, 2021, 10 (1): e1118.

[267] Wang Y N, Xu R, Wang H, et al. Insights into the stabilization of landfill by assessing the diversity and dynamic succession of bacterial community and its associated bio-metabolic process [J]. Science of the Total Environment, 2021 (768): 145466.

[268] Rodrigues G R, Pinto O H B, Schroeder L F, et al. Unraveling the xylanolytic potential of *Acidobacteria bacterium* AB60 from Cerrado soils [J]. FEMS Microbiol Lett, 2020, 367 (18): 149.

[269] Gutierrez T, Green D H, Nichols P D, et al. *Polycyclovorans algicola* gen. nov., sp. nov., an aromatic-hydrocarbon-degrading marine bacterium found associated with laboratory cultures of marine phytoplankton [J]. Applied and Environmental Microbiology, 2012, 79 (1): 205-214.

[270] Gutierrez T, Thompson H F, Angelova A, et al. Genome sequence of *Polycyclovorans algicola* strain TG408, an obligate polycyclic aromatic hydrocarbon-degrading bacterium associated with marine eukaryotic phytoplankton [J]. Genome Announcements, 2015, 3 (2): e00207-15.

[271] Kumar S, Suyal D C, Bhoriyal M, et al. Plant growth promoting potential of psychrotolerant *Dyadobacter* sp for pulses and finger millet and impact of inoculation on soil chemical properties and diazotrophic abun-

dance [J]. Journal of Plant Nutrition, 2018, 41 (8): 1035-1046.
[272] Yadav S, Khan M A, Sharma R, et al. Potential of formulated *Dyadobacter jiangsuensis* strain 12851 for enhanced bioremediation of chlorpyrifos contaminated soil [J]. Ecotoxicology and Environmental Safety, 2021 (213): 112039.
[273] Zhu F X, Zhu C Y, Doyle E, et al. Fate of di (2-ethylhexyl) phthalate in different soils and associated bacterial community changes [J]. Science of the Total Environment, 2018 (637): 460-469.
[274] Rezgui R, Maaroufi A, Fardeau M L, et al. *Anaerosalibacter bizertensis* gen. nov., sp. nov., a halotolerant bacterium isolated from sludge [J]. International Journal of Systematic and Evolutionary Microbiology, 2012, 62 (10): 2469-2474.
[275] Zhang Q, Zhu L, Wang J, et al. Effects of fomesafen on soil enzyme activity, microbial population, and bacterial community composition [J]. Environmental Monitoring and Assessment, 2014, 186 (5): 2801-2812.
[276] Hou F, Du J J, Bi X, et al. Toxicity effects of aged refuse on *Tagetes patula* and rhizosphere microbes [J]. Land Degradation & Development, 2022, 33: 1043-1053.
[277] Verma S K, Das A K, Gantait S, et al. Applications of carbon nanomaterials in the plant system: a perspective view on the pros and cons [J]. Science of the Total Environment, 2019 (667): 485-499.
[278] Tong Z H, Bischoff M, Nies L, et al. Impact of fullerene (C_{60}) on a soil microbial community [J]. Environmental Science & Technology, 2007, 41 (8): 2985-2991.
[279] Du J J, Wang T, Zhou Q X, et al. Graphene oxide enters the rice roots and disturbs the endophytic bacterial communities [J]. Ecotoxicology and Environmental Safety, 2020 (192): 110304.
[280] Zhao J, Wang Z Y, White J C, et al. Graphene in the aquatic environment: adsorption, dispersion, toxicity and transformation [J]. Environmental Science & Technology, 2014, 48 (17): 9995-10009.
[281] Anjum N A, Singh N, Singh M K, et al. Single-bilayer graphene oxide

sheet impacts and underlying potential mechanism assessment in germinating faba bean (*Vicia faba* L.) [J]. Science of the Total Environment, 2014 (472): 834-841.

[282] Etesami H, Hosseini H M, Alikhani H A. In planta selection of plant growth promoting endophytic bacteria for rice (*Oryza sativa* L.) [J]. Journal of Soil Science and Plant Nutrition, 2014, 14 (2): 491-503.

[283] Mano H, Morisaki H. Endophytic bacteria in the rice plant [J]. Microbes and Environments, 2008, 23 (2): 109-117.

[284] Etesami H, Hosseini H M, Alikhani H A. In planta selection of plant growth promoting endophytic bacteria for rice (*Oryza sativa* L.) [J]. Journal of Soil Science & Plant Nutrition, 2014 (14): 491-503.

[285] Feng Y, Shen D, Song W. Rice endophyte *Pantoea agglomerans* YS19 promotes host plant growth and affects allocations of host photosynthates [J]. Journal of Applied Microbiology, 2010, 100 (5): 938-945.

[286] Feng F Y, Ge J, Li Y S, et al. Isolation, colonization, and chlorpyrifos degradation mediation of the endophytic bacterium *Sphingomonas* Strain HJY in Chinese Chives (*Allium tuberosum*) [J]. Journal of Agricultural and Food Chemistry, 2017, 65 (6): 1131-1138.

[287] Suyamud B, Thiravetyan P, Panyapinyopol B, et al. Dracaena sanderiana endophytic bacteria interactions: effect of endophyte inoculation on bisphenol a removal [J]. Ecotoxicology and Environmental Safety, 2018 (157): 318-326.

[288] Sun L, Qiu F, Zhang X, et al. Endophytic bacterial eiversity in rice (*Oryza sativa* L.) roots estimated by 16S rDNA sequence analysis [J]. Microbial Ecology, 2008, 55 (3): 415-424.

[289] Song G L, Gao Y, Wu H, et al. Physiological effect of anatase TiO_2 nanoparticles on *Lemna minor* [J]. Environmental Toxicology and Chemistry, 2012, 31 (9): 2147-2152.

[290] Tian X Y, Zhang C S. Illumina-based analysis of endophytic and rhizosphere bacterial diversity of the coastal halophyte messerschmidia sibirica [J]. Frontiers in Microbiology, 2017 (8): 2288-2288.

[291] Amato K R, Yeoman C J, Kent A, et al. Habitat degradation impacts

black howler monkey (*Alouatta pigra*) gastrointestinal microbiomes [J]. Isme Journal, 2013, 7 (7): 1344-1353.

[292] Omid A, Elham G. Toxicity of graphene and graphene oxide nanowalls against bacteria [J]. ACS Nano, 2010, 4 (10): 5731-5736.

[293] Khodakovskaya M, Dervishi E, Mahmood M, et al. Carbon nanotubes are able to penetrate plant seed coat and dramatically affect seed germination and plant growth [J]. ACS Nano, 2009, 3 (10): 3221-3227.

[294] Zhang P, Zhang R, Fang X, et al. Toxic effects of graphene on the growth and nutritional levels of wheat (*Triticum aestivum* L.): short- and long-term exposure studies [J]. Journal of Hazardous Materials, 2016, 317 (5): 543-551.

[295] Hu X, Jia K, Lu K, et al. Graphene oxide amplifies the phytotoxicity of arsenic in wheat [J]. Scientific Reports, 2014 (4): 6122.

[296] Li Y, Yuan H, Von dem Bussche A, et al. Graphene microsheets enter cells through spontaneous membrane penetration at edge asperities and corner sites [J]. Proceedings of the National Academy of Sciences of the United States of America, 2013, 110 (30): 12295-12300.

[297] Hui L, Ye D, Wang X, et al. Soil bacterial communities of different natural forest types in Northeast China [J]. Plant and Soil, 2014, 383 (1-2): 203-216.

[298] Huang J, Wang Z W, Zhu C W, et al. Identification of microbial communities in open and closed circuit bioelectrochemical MBRs by high-throughput 454 pyrosequencing [J]. Plos One, 2014, 9 (4): e93842.

[299] Chaudhary H J, Peng G X, Hu M, et al. Genetic diversity of endophytic diazotrophs of the wild rice, *Oryza alta* and identification of the new diazotroph, *Acinetobacter oryzae* sp. nov [J]. Microbial Ecology, 2012, 63 (4): 813-821.

[300] Visioli G, Vamerali T, Mattarozzi M, et al. Combined endophytic inoculants enhance nickel phytoextraction from serpentine soil in the hyperaccumulator *Noccaea caerulescens* [J]. Frontiers in Plant Science, 2015, 6: 638.

[301] Chimwamurombe P M, Gronemeyer J L, Reinhold-Hurek B. Isolation

and characterization of culturable seed-associated bacterial endophytes from gnotobiotically grown *Marama* bean seedlings [J]. Fems Microbiology Ecology, 2016, 92 (6): fiw083.

[302] Loaces I, Ferrando L, Scavino A F. Dynamics, diversity and function of endophytic siderophore-producing bacteria in rice [J]. Microbial Ecology, 2011, 61 (3): 606-618.

[303] Idris R, Trifonova R, Puschenreiter M, et al. Bacterial communities associated with flowering plants of the Ni hyperaccumulator *Thlaspi goesingense* [J]. Applied and Environmental Microbiology, 2004, 70 (5): 2667-2677.

[304] Martinez-Rodriguez J D, De la Mora-Amutio M, Plascencia-Correa L A, et al. Cultivable endophytic bacteria from leaf bases of *Agave tequilana* and their role as plant growth promoters [J]. Brazilian Journal of Microbiology, 2014, 45 (4): 1333-1339.

[305] Wang Y Y, Yang X E, Zhang X C, et al. Improved plant growth and Zn accumulation in grains of rice (*Oryza sativa* L.) by inoculation of endophytic microbes isolated from a Zn hyperaccumulator, *Sedum alfredii* H [J]. Journal of Agricultural and Food Chemistry, 2014, 62 (8): 1783-1791.

[306] Pereira S I A, Castro P M L. Diversity and characterization of culturable bacterial endophytes from *Zea mays* and their potential as plant growth-promoting agents in metal-degraded soils [J]. Environmental Science and Pollution Research, 2014, 21 (24): 14110-14123.

[307] Bacilio-Jimenez M, Aguilar-Flores S, del Valle M V, et al. Endophytic bacteria in rice seeds inhibit early colonization of roots by *Azospirillum brasilense* [J]. Soil Biology and Biochemistry, 2001, 33 (2): 167-172.

[308] Akinsanya M A, Goh J K, Lim S P, et al. Diversity, antimicrobial and antioxidant activities of culturable bacterial endophyte communities in *Aloe vera* [J]. Fems Microbiology Letters, 2015, 362 (23): fnv184.

[309] Liu Y, Wang R H, Cao Y H, et al. Identification and antagonistic activity of endophytic bacterial strain *Paenibacillus* sp 5 L8 isolated from the seeds of maize (*Zea mays* L., Jingke 968) [J]. Annals of Microbi-

ology, 2016, 66 (2): 653-660.
[310] Liu S B, Zeng T H, Hofmann M, et al. Antibacterial activity of graphite, graphite oxide, graphene oxide, and reduced graphene oxide: membrane and oxidative stress [J]. ACS Nano, 2011, 5 (9): 6971-6980.
[311] Hu W B, Peng C, Luo W J, et al. Graphene-based antibacterial paper [J]. ACS Nano, 2010, 4 (7): 4317-4323.
[312] Akhavan O, Ghaderi E. Toxicity of graphene and graphene oxide nanowalls against bacteria [J]. Acs Nano, 2010, 4 (10): 5731-5736.
[313] Dubois-Brissonnet F, Trotier E, Briandet R. The biofilm life style involves an increase in bacterial membrane saturated fatty acids [J]. Front Microbiol, 2016, (7): 1673.
[314] Joshi R, Carbone P, Wang F, et al. Precise and ultrafast molecular sieving through graphene oxide membranes [J]. Science, 2014, 343 (6172): 752-754.
[315] Dreyer D R, Park S, Bielawski C W, et al. The chemistry of graphene oxide [J]. Chemical Society Reviews, 2010, 39 (1): 228-40.
[316] Ahmed F, Rodrigues D F. Investigation of acute effects of graphene oxide on wastewater microbial community: a case study [J]. Journal of Hazardous Materials, 2013 (256): 33-39.
[317] Zhao J, Wang Z, White J C, et al. Graphene in the aquatic environment: adsorption, dispersion, toxicity and transformation [J]. Environmental Science & Technology, 2014, 48 (17): 9995-10009.
[318] Anjum N A, Singh N, Singh M K, et al. Single-bilayer graphene oxide sheet tolerance and glutathione redox system significance assessment in faba bean (*Vicia faba* L.) [J]. Journal of Nanoparticle Research, 2013, 15 (7): 1-12.
[319] Anjum N A, Singh N, Singh M K, et al. Single-bilayer graphene oxide sheet impacts and underlying potential mechanism assessment in germinating faba bean (*Vicia faba* L.) [J]. Science of the Total Environment, 2014 (472): 834-841.
[320] Nair R, Mohamed M S, Gao W, et al. Effect of carbon nanomaterials on the germination and growth of rice plants [J]. Journal of Nanoscience

and Nanotechnology, 2012, 12 (3): 2212-2220.
[321] Qu X, Brame J, Li Q, et al. Nanotechnology for a safe and sustainable water supply: enabling integrated water treatment and reuse [J]. Accounts of Chemical Research, 2012, 46 (3): 834-843.
[322] Alvarez P J, Colvin V, Lead J, et al. Research priorities to advance eco-responsible nanotechnology [J]. ACS Nano, 2009, 3 (7): 1616-1619.
[323] Lee J, Mahendra S, Alvarez P J. Nanomaterials in the construction industry: a review of their applications and environmental health and safety considerations [J]. ACS Nano, 2010, 4 (7): 3580-3590.
[324] Hu X, Zhou Q. Novel hydrated graphene ribbon unexpectedly promotes aged seed germination and root differentiation [J]. Scientific Reports, 2014, 4 (1): 3782.
[325] Hu X, Mu L, Kang J, et al. Humic acid acts as a natural antidote of graphene by regulating nanomaterial translocation and metabolic fluxes in vivo [J]. Environmental Science & Technology, 2014, 48 (12): 6919-6927.
[326] Kuila T, Bose S, Khanra P, et al. A green approach for the reduction of graphene oxide by wild carrot root [J]. Carbon, 2012, 50 (3): 914-921.
[327] Akhavan O, Ghaderi E. *Escherichia coli* bacteria reduce graphene oxide to bactericidal graphene in a self-limiting manner [J]. Carbon, 2012, 50 (5): 1853-1860.
[328] Wang G, Qian F, Saltikov C W, et al. Microbial reduction of graphene oxide by *Shewanella* [J]. Nano Research, 2011, 4 (6): 563-570.
[329] Salas E C, Sun Z, Luttge A, et al. Reduction of graphene oxide via bacterial respiration [J]. ACS nano, 2010, 4 (8): 4852-4856.
[330] Zhu C, Feng Z, Fan M, et al. Biosynthesis approach to nitrogen doped graphene by denitrifying bacteria CFMI-1 [J]. RSC Advances, 2014, 4 (76): 40292-40295.
[331] Holden P A, Nisbet R M, Lenihan H S, et al. Ecological nanotoxicology: integrating nanomaterial hazard considerations across the subcellular, population, community, and ecosystems levels [J]. Accounts of Chemical Research, 2012, 46 (3): 813-822.

参 考 文 献

[332] Holden P A, Klaessig F, Turco R F, et al. Evaluation of exposure concentrations used in assessing manufactured nanomaterial environmental hazards: are they relevant [J]. Environmental Science & Technology, 2014, 48 (18): 10541-10551.

[333] Rico C M, Majumdar S, Duarte-Gardea M, et al. Interaction of nanoparticles with edible plants and their possible implications in the food chain [J]. Journal of Agricultural and Food Chemistry, 2011, 59 (8): 3485-3498.

[334] Lyubenova L, Kuhn A J, Höltkemeier A, et al. Root exudation pattern of *Typha latifolia* L. plants after copper exposure [J]. Plant and Soil, 2013, 370 (1-2): 187-195.

[335] Quartacci M F, Irtelli B, Gonnelli C, et al. Naturally-assisted metal phytoextraction by *Brassica carinata*: role of root exudates [J]. Environmental Pollution, 2009, 157 (10): 2697-2703.

[336] Bergqvist C, Herbert R, Persson I, et al. Plants influence on arsenic availability and speciation in the rhizosphere, roots and shoots of three different vegetables [J]. Environmental Pollution, 2014 (184): 540-546.

[337] LeFevre G H, Hozalski R M, Novak P J. Root exudate enhanced contaminant desorption: an abiotic contribution to the rhizosphere effect [J]. Environmental Science & Technology, 2013, 47 (20): 11545-11553.

[338] Huang Y C, Fan R, Grusak M A, et al. Effects of nano-ZnO on the agronomically relevant *Rhizobium legume* symbiosis [J]. Science of the Total Environment, 2014 (497): 78-90.

[339] Du J J, Hu X G, Mu L, et al. Root exudates as natural ligands that alter the properties of graphene oxide and environmental implications thereof [J]. RSC Advances, 2015, 5 (23): 17615-17622.

[340] Baetz U, Martinoia E. Root exudates: the hidden part of plant defense [J]. Trends in Plant Science, 2014, 19 (2): 90-98.

[341] Kiser M A, Ladner D A, Hristovski K D, et al. Nanomaterial transformation and association with fresh and freeze-dried wastewater activated sludge: implications for testing protocol and environmental fate [J]. En-

vironmental Science & Technology, 2012, 46 (13): 7046-7053.

[342] Kim S, Zhou S, Hu Y, et al. Room-temperature metastability of multilayer graphene oxide films [J]. Nature Materials, 2012, 11 (6): 544-549.

[343] Smith S C, Ahmed F, Gutierrez K M, et al. A comparative study of lysozyme adsorption with graphene, graphene oxide, and single-Walled carbon nanotubes: potential environmental applications [J]. Chemical Engineering Journal, 2014 (240): 147-154.

[344] Beless B, Rifai H S, Rodrigues D F. Efficacy of carbonaceous materials for sorbing polychlorinated biphenyls from aqueous solution [J]. Environmental Science & Technology, 2014, 48 (17): 10372-10379.

[345] Fan W, Shi Z, Yang X, et al. Bioaccumulation and biomarker responses of cubic and octahedral Cu_2O micro/nanocrystals in *Daphnia magna* [J]. Water Research, 2012, 46 (18): 5981-5988.

[346] Martinolich A J, Park G, Nakamoto M Y, et al. Structural and functional effects of Cu metalloprotein-driven silver nanoparticle dissolution [J]. Environmental Science & Technology, 2012, 46 (11): 6355-6362.

[347] Zhang L, Li X, Huang Y, et al. Controlled synthesis of few-layered graphene sheets on a large scale using chemical exfoliation [J]. Carbon, 2010, 48 (8): 2367-2371.

[348] Wu L, Zhang Y, Zhang C, et al. Tuning cell autophagy by diversifying carbon nanotube surface chemistry [J]. ACS Nano, 2014, 8 (3): 2087-2099.

[349] Liao K H, Lin Y S, Macosko C W, et al. Cytotoxicity of graphene oxide and graphene in human erythrocytes and skin fibroblasts [J]. ACS Applied Materials & Interfaces, 2011, 3 (7): 2607-2615.

[350] Shomer I, Novacky A J, Pike S M, et al. Electrical potentials of plant cell walls in response to the ionic environment [J]. Plant Physiology, 2003, 133 (1): 411-422.

[351] Dimiev A M, Tour J M. Mechanism of Graphene Oxide Formation [J]. ACS Nano, 2014, 8 (3): 3060-3068.

[352] Sheng Z H, Shao L, Chen J J, et al. Catalyst-free synthesis of nitrogen-

doped graphene via thermal annealing graphite oxide with melamine and its excellent electrocatalysis [J]. ACS Nano, 2011, 5 (6): 4350-5358.

[353] Yang Z, Yao Z, Li G, et al. Sulfur-doped graphene as an efficient metal-free cathode catalyst for oxygen reduction [J]. ACS nano, 2011, 6 (1): 205-211.

[354] Fischer S, Papageorgiou A C, Marschall M, et al. L-cysteine on Ag (111): a combined STM and X-ray spectroscopy study of anchorage and deprotonation [J]. The Journal of Physical Chemistry C, 2012, 116 (38): 20356-20362.

[355] Yang K, Li Y, Tan X, et al. Behavior and toxicity of graphene and its functionalized derivatives in biological systems [J]. Small, 2013, 9 (9-10): 1492-1503.

[356] George S, Lin S, Ji Z, et al. Surface defects on plate-shaped silver nanoparticles contribute to its hazard potential in a fish gill cell line and zebrafish embryos [J]. ACS Nano, 2012, 6 (5): 3745-3759.

[357] Rao S, Jammalamadaka S N, Stesmans A, et al. Ferromagnetism in graphene nanoribbons: split versus oxidative unzipped ribbons [J]. Nano Letters, 2012, 12 (3): 1210-1217.

[358] Hu X, Mu L, Lu K, et al. Green synthesis of low-toxicity graphene-fulvic acid with an open band gap enhances demethylation of methylmercury [J]. ACS Applied Materials & Interfaces, 2014, 6 (12): 9220-9227.

[359] Yin H, Tang H, Wang D, et al. Facile synthesis of surfactant-free Au cluster/graphene hybrids for high-performance oxygen reduction reaction [J]. ACS Nano, 2012, 6 (9): 8288-8297.

[360] Luo Z, Lu Y, Somers L A, et al. High yield preparation of macroscopic graphene oxide membranes [J]. Journal of the American Chemical Society, 2009, 131 (3): 898-899.

[361] Zhu C, Guo S, Fang Y, et al. Reducing sugar: new functional molecules for the green synthesis of graphene nanosheets [J]. ACS nano, 2010, 4 (4): 2429-2437.

[362] Muller J, Huaux F, Fonseca A, et al. Structural defects play a major role in the acute lung toxicity of multiwall carbon nanotubes: toxicologi-

cal aspects [J]. Chemical Research in Toxicology, 2008, 21 (9): 1698-1705.

[363] Wang Q, Lee S, Choi H. Aging study on the structure of Fe0-nanoparticles: stabilization, characterization, and reactivity [J]. The Journal of Physical Chemistry C, 2010, 114 (5): 2027-2033.

[364] Fenoglio I, Greco G, Tomatis M, et al. Structural defects play a major role in the acute lung toxicity of multiwall carbon nanotubes: physicochemical aspects [J]. Chemical Research in Toxicology, 2008, 21 (9): 1690-1697.

[365] Stensberg M C, Madangopal R, Yale G, et al. Silver nanoparticle-specific mitotoxicity in *Daphnia magna* [J]. Nanotoxicology, 2014, 8 (8): 833-842.

[366] Buerki-Thurnherr T, Xiao L, Diener L, et al. In vitro mechanistic study towards a better understanding of ZnO nanoparticle toxicity [J]. Nanotoxicology, 2013, 7 (4): 402-416.

[367] Labille J, Feng J, Botta C, et al. Aging of TiO_2 nanocomposites used in sunscreen. dispersion and fate of the degradation products in aqueous environment [J]. Environmental Pollution, 2010, 158 (12): 3482-3489.

[368] 杜俊杰, 李娜, 吴建虎. 不同纳米材料对小麦种子萌发的影响[J]. 安徽农业科学, 2018, 46 (13): 38-40, 124.

[369] Thuesombat P, Hannongbua S, Akasit S, et al. Effect of silver nanoparticles on rice (*Oryza sativa* L. cv. KDML 105) seed germination and seedling growth [J]. Ecotoxicology and Environmental Safety, 2014 (104): 302-309.

[370] Pittol M, Tomacheski D, Simões D N, et al. Macroscopic effects of silver nanoparticles and titanium dioxide on edible plant growth [J]. Environmental Nanotechnology, Monitoring & Management, 2017 (8): 127-133.

[371] Barabanov P V, Gerasimov A V, Blinov A V, et al. Influence of nanosilver on the efficiency of *Pisum sativum* crops germination [J]. Ecotoxicology and Environmental Safety, 2018 (147): 715-719.

[372] Mariya K, Enkeleda D, Meena M, et al. Carbon nanotubes are able to

penetrate plant seed coat and dramatically affect seed germination and plant growth [J]. ACS Nano, 2009, 3 (10): 3221-3227.

[373] 姜余梅, 刘强, 赵怡情, 等. 碳纳米管对水稻种子萌发和根系生长的影响[J]. 湖北农业科学, 2014 (5): 1010-1012.

[374] Begum P, Ikhtiari R, Fugetsu B, et al. Phytotoxicity of multi-walled carbon nanotubes assessed by selected plant species in the seedling stage [J]. Applied Surface Science, 2012 (262): 120-124.

[375] Yan S, Zhao L, Li H, et al. Single-walled carbon nanotubes selectively influence maize root tissue development accompanied by the change in the related gene expression [J]. Journal of Hazardous Materials, 2013 (246-247): 110-118.

[376] Zhang P, Zhang R, Fang X, et al. Toxic effects of graphene on the growth and nutritional levels of wheat (*Triticum aestivum* L.): short- and long-term exposure studies [J]. Journal of Hazardous Materials, 2016 (317): 543-551.

[377] 吴金海, 焦靖芝, 谢伶俐, 等. 氧化石墨烯处理对甘蓝型油菜生长发育的影响[J]. 基因组学与应用生物学, 2015 (12): 2738-2742.

[378] Begum P, Ikhtiari R, Fugetsu B. Graphene phytotoxicity in the seedling stage of cabbage, tomato, red spinach, and lettuce [J]. Carbon, 2011, 49 (12): 3907-3919.

[379] Dikin D A, Stankovich S, Zimney E J, et al. Preparation and characterization of graphene oxide paper [J]. Nature, 2007, 448 (7152): 457-460.

[380] Qu X, Alvarez P J, Li Q. Applications of nanotechnology in water and wastewater treatment [J]. Water Research, 2013, 47 (12): 3931-3946.

[381] Sanchez V C, Jachak A, Hurt R H, et al. Biological interactions of graphene-family nanomaterials: an interdisciplinary review [J]. Chemical Research in Toxicology, 2011, 25 (1): 15-34.

[382] Shrestha B, Acosta-Martinez V, Cox S B, et al. An evaluation of the impact of multiwalled carbon nanotubes on soil microbial community structure and functioning [J]. Journal of Hazardous Materials, 2013

(261): 188-197.

[383] Nowack B, Bucheli T D. Occurrence, behavior and effects of nanoparticles in the environment [J]. Environmental Pollution, 2007, 150 (1): 5-22.

[384] Dinesh R, Anandaraj M, Srinivasan V, et al. Engineered nanoparticles in the soil and their potential implications to microbial activity [J]. Geoderma, 2012 (173): 19-27.

[385] Zhao J, Liu F, Wang Z, et al. Heteroaggregation of graphene oxide with minerals in aqueous phase [J]. Environmental Science & Technology, 2015, 49 (5): 2849-2857.

[386] Tong Z H, Bischoff M, Nies L F, et al. Response of soil microorganisms to as-produced and functionalized single-wall carbon nanotubes (SWNTs) [J]. Environmental Science & Technology, 2012, 46 (24): 13471-13479.

[387] Chung H, Son Y, Yoon T K, et al. The effect of multi-walled carbon nanotubes on soil microbial activity [J]. Ecotoxicology and Environmental Safety, 2011, 74 (4): 569-575.

[388] Liu S, Zeng T H, Hofmann M, et al. Antibacterial activity of graphite, graphite oxide, graphene oxide, and reduced graphene oxide: membrane and oxidative stress [J]. ACS Nano, 2011, 5 (9): 6971-6980.

[389] Carpio I E M, Santos C M, Wei X, et al. Toxicity of a polymer-graphene oxide composite against bacterial planktonic cells, biofilms, and mammalian cells [J]. Nanoscale, 2012, 4 (15): 4746-4756.

[390] Kang S, Mauter M S, Elimelech M. Microbial cytotoxicity of carbon-based nanomaterials: implications for river water and wastewater effluent [J]. Environmental Science & Technology, 2009, 43 (7): 2648-2653.

[391] Khanra P, Kuila T, Kim N H, et al. Simultaneous bio-functionalization and reduction of graphene oxide by baker's yeast [J]. Chemical Engineering Journal, 2012 (183): 526-533.

[392] Rodrigues D F, Jaisi D P, Elimelech M. Toxicity of functionalized single-walled carbon nanotubes on soil microbial communities: implications for nutrient cycling in soil [J]. Environmental Science & Technology,

2013, 47 (1): 625-633.

[393] Du J J, Hu X G, Zhou Q X. Reply to the 'Comment on "Graphene oxide regulates the changes in soil"' by C. Forstner, P. Wang, P. M. Kopittke and P. G. Dennis, RSC Adv., 2016, 6, DOI: 10. 1039/C5RA26329H [J]. RSC Advances, 2016, 6 (59): 53688-53689.

[394] Jiao Y, Qian F, Li Y, et al. Deciphering the electron transport pathway for graphene oxide reduction by *Shewanella oneidensis* MR-1 [J]. Journal of Bacteriology, 2011, 193 (14): 3662-2665.

[395] Lee K R, Lee K U, Lee J W, et al. Electrochemical oxygen reduction on nitrogen doped graphene sheets in acid media [J]. Electrochemistry Communications, 2010, 12 (8): 1052-1055.

[396] Su Y, Hu M, Fan C, et al. The cytotoxicity of CdTe quantum dots and the relative contributions from released cadmium ions and nanoparticle properties [J]. Biomaterials, 2010, 31 (18): 4829-4834.

[397] Kotchey G P, Allen B L, Vedala H, et al. The enzymatic oxidation of graphene oxide [J]. ACS Nano, 2011, 5 (3): 2098-2108.

[398] Lahiani M H, Dervishi E, Chen J, et al. Impact of carbon nanotube exposure to seeds of valuable crops [J]. ACS Applied Materials & Interfaces, 2013, 5 (16): 7965-7973.

[399] Fen Hou, Tian Li, Xu Bi, et al. Effects of aged refuse on endogenous hormone contents in leaves of *Tagetes patula* L [J]. Fresenius Environmental Bulletin, 2022.

[400] Wang J, Chen Z M, Chen B L. Adsorption of polycyclic aromatic hydrocarbons by graphene and graphene oxide nanosheets [J]. Environmental Science & Technology, 2014, 48 (9): 4817-4825.

[401] Pulizzi F. Nano in the future of crops [J]. Nature Nanotechnology, 2019, 14 (6): 507.

[402] Yin J, Wang Y, Gilbertson L M. Opportunities to advance sustainable design of nano-enabled agriculture identified through a literature review [J]. Environmental Science: Nano, 2018, 5 (1): 11-26.

[403] Kah M, Tufenkji N, White J C. Nano-enabled strategies to enhance crop nutrition and protection [J]. Nature Nanotechnology, 2019, 14 (6):

532-540.

[404] Ren X Y, Zeng G M, Tang L, et al. Effect of exogenous carbonaceous materials on the bioavailability of organic pollutants and their ecological risks [J]. Soil Biology and Biochemistry, 2018 (116): 70-81.

[405] Gibbons S M. Microbial community ecology: function over phylogeny [J]. Nature Ecology & Evolution, 2017, 1 (1): 32.

[406] Ge Y, Shen C C, Wang Y, et al. Carbonaceous nanomaterials have higher effects on soybean rhizosphere prokaryotic communities during the reproductive growth phase than during vegetative growth [J]. Environmental Science & Technology, 2018, 52 (11): 6636-6646.

[407] Mu L, Zhou Q X, Zhao Y J, et al. Graphene oxide quantum dots stimulate indigenous bacteria to remove oil contamination [J]. Journal of Hazardous Materials, 2019 (366): 694-702.

[408] Liu L Y, Wang J Z, Wei G L, et al. Sediment records of polycyclic aromatic hydrocarbons (PAHs) in the continental shelf of China: Implications for evolving anthropogenic impacts [J]. Environmental Science & Technology, 2012, 46 (12): 6497-6504.

[409] Song M, Luo C L, Jiang L F, et al. Identification of benzo a pyrene-metabolizing bacteria in forest soils by using DNA-based stable-isotope probing [J]. Applied and Environmental Microbiology, 2015, 81 (21): 7368-7376.

[410] Saeedi M, Li L Y, Salmanzadeh M. Heavy metals and polycyclic aromatic hydrocarbons: Pollution and ecological risk assessment in street dust of *Tehran* [J]. Journal of Hazardous Materials, 2012 (227): 9-17.

[411] Kim K H, Jahan S A, Kabir E, et al. A review of airborne polycyclic aromatic hydrocarbons (PAHs) and their human health effects [J]. Environment International, 2013 (60): 71-80.

[412] Li Q, Kim M, Liu Y, et al. Quantitative assessment of human health risks induced by vehicle exhaust polycyclic aromatic hydrocarbons at Zhengzhou via multimedia fugacity models with cancer risk assessment [J]. Science of the Total Environment, 2018 (618): 430-438.

[413] Wang J, Liu J, Ling W T, et al. Composite of PAH-degrading endo-

phytic bacteria reduces contamination and health risks caused by PAHs in vegetables [J]. Science of the Total Environment, 2017 (598): 471-478.

[414] Gateuille D, Evrard O, Lefevre I, et al. Mass balance and decontamination times of polycyclic aromatic hydrocarbons in rural nested catchments of an early industrialized region (Seine River basin, France) [J]. Science of the Total Environment, 2014 (470): 608-617.

[415] Ren G, Ren W J, Teng Y, et al. Evident bacterial community changes but only slight degradation when polluted with pyrene in a red soil [J]. Frontiers in Microbiology, 2015 (6): 22.

[416] Cristaldi A, Conti G O, Jho E H, et al. Phytoremediation of contaminated soils by heavy metals and PAHs: a brief review [J]. Environmental Technology & Innovation, 2017 (8): 309-326.

[417] Sun T R, Cang L, Wang Q Y, et al. Roles of abiotic losses, microbes, plant roots, and root exudates on phytoremediation of PAHs in a barren soil [J]. Journal of Hazardous Materials, 2010, 176 (1-3): 919-925.

[418] Chen X, Liu X Y, Zhang X Y, et al. Phytoremediation effect of *Scirpus triqueter* noculated plant-growth-promoting bacteria (PGPB) on different fractions of pyrene and Ni in co-contaminated soils [J]. Journal of Hazardous Materials, 2017 (325): 319-326.

[419] Sun Y B, Zhou Q X, Xu Y M, et al. Phytoremediation for co-contaminated soils of benzo a pyrene (BaP) and heavy metals using ornamental plant *Tagetes patula* [J]. Journal of Hazardous Materials, 2011, 186 (2-3): 2075-2082.

[420] Li H, Li X, Xiang L, et al. Phytoremediation of soil co-contaminated with Cd and BDE-209 using hyperaccumulator enhanced by AM fungi and surfactant [J]. Science of the Total Environment, 2018 (613): 447-455.

[421] Chowdhury I, Duch M C, Mansukhani N D, et al. Colloidal properties and stability of graphene oxide nanomaterials in the aquatic environment [J]. Environmental Science & Technology, 2013, 47 (12): 6288-6296.

[422] Singh S K, Singh M K, Kulkarni P P, et al. Amine-modified graphene. thrombo-protective safer alternative to graphene oxide for biomedical applications [J]. ACS Nano, 2012, 6 (3): 2731-2740.

[423] Ren W C, Cheng H M. The global growth of graphene [J]. Nature Nanotechnology, 2014, 9 (10): 726-730.

[424] Xiong T, Yuan X Z, Wang H, et al. Implication of graphene oxide in Cd-contaminated soil: a case study of bacterial communities [J]. Journal of Environmental Management, 2018 (205): 99-106.

[425] Du J, Zhou Q, Wu J, et al. Soil bacterial communities respond differently to graphene oxide and reduced graphene oxide after 90 days of exposure [J]. Soil Ecology Letters, 2020, 2 (3): 176-179.

[426] He Y J, Hu R R, Zhong Y J, et al. Graphene oxide as a water transporter promoting germination of plants in soil [J]. Nano Research, 2018, 11 (4): 1928-1937.

[427] 孙约兵, 周启星, 任丽萍. 一种利用孔雀草治理并修复苯并 [a] 芘污染土壤的方法: CN101992206A [P]. 2011-03-30.

[428] Sun Y, Zhou Q, Xu Y, et al. Phytoremediation for co-contaminated soils of benzo [a] pyrene (B [a] P) and heavy metals using ornamental plant *Tagetes patula* [J]. Journal of Hazardous Materials, 2011, 186 (2): 2075-2082.

[429] Greenwood S J, Rutter A, Zeeb B A. The absorption and translocation of polychlorinated biphenyl congeners by *Cucurbita pepo* ssp *pepo* [J]. Environmental Science & Technology, 2011, 45 (15): 6511-6.

[430] Singer A C, Smith D, Jury W A, et al. Impact of the plant rhizosphere and augmentation on remediation of polychlorinated biphenyl contaminated soil [J]. Environmental Toxicology and Chemistry, 2003, 22 (9): 1998-2004.

[431] Shen C F, Tang X J, Cheema S A, et al. Enhanced phytoremediation potential of polychlorinated biphenyl contaminated soil from e-waste recycling area in the presence of randomly methylated-beta-cyclodextrins [J]. Journal of Hazardous Materials, 2009, 172 (2-3): 1671-1676.

[432] Lin Q, Shen K L, Zhao H M, et al. Growth response of *Zea mays* L. in pyrene-copper co-contaminated soil and the fate of pollutants [J]. Journal of Hazardous Materials, 2008, 150 (3): 515-521.

[433] Yu X Z, Wu S C, Wu F Y, et al. Enhanced dissipation of PAHs from

soil using mycorrhizal ryegrass and PAH-degrading bacteria [J]. Journal of Hazardous Materials, 2011, 186 (2-3): 1206-1217.

[434] Abhilash P C, Powell J R, Singh H B, et al. Plant-microbe interactions: novel applications for exploitation in multipurpose remediation technologies [J]. Trends in Biotechnology, 2012, 30 (8): 416-420.

[435] Yoshitomi K J, Shann J R. Corn (*Zea mays* L.) root exudates and their impact on C-14-pyrene mineralization [J]. Soil Biology and Biochemistry, 2001, 33 (12-13): 1769-1776.

[436] Qi Z C, Hou L, Zhu D Q, et al. Enhanced transport of phenanthrene and 1-naphthol by colloidal graphene oxide nanoparticles in saturated soil [J]. Environmental Science & Technology, 2014, 48 (17): 10136-10144.

[437] Ren X Y, Zeng G M, Tang L, et al. Effect of exogenous carbonaceous materials on the bioavailability of organic pollutants and their ecological risks [J]. Soil Biology & Biochemistry, 2018 (116): 70-81.

[438] Su Y L, Zheng X, Chen A H, et al. Hydroxyl functionalization of single-walled carbon nanotubes causes inhibition to the bacterial denitrification process [J]. Chemical Engineering Journal, 2015 (279): 47-55.

[439] Rajavel K, Gomathi R, Manian S, et al. In vitro bacterial cytotoxicity of CNTs: reactive oxygen species mediate cell damage edges over direct physical puncturing [J]. Langmuir, 2014, 30 (2): 592-601.

[440] Zhu B T, Wu S, Xia X H, et al. Effects of carbonaceous materials on microbial bioavailability of 2, 2', 4, 4'-tetrabromodiphenyl ether (BDE-47) in sediments [J]. Journal of Hazardous Materials, 2016, (312): 216-223.

[441] Rong Y, Yang Y, Guan Y N, et al. Pyrosequencing reveals soil enzyme activities and bacterial communities impacted by graphene and its oxides [J]. Journal of Agricultural and Food Chemistry, 2017, 65 (42): 9191-9199.

[442] Khodakovskaya M V, Kim B S, Kim J N, et al. Carbon nanotubes as plant growth regulators: Effects on tomato growth, reproductive system, and soil microbial community [J]. Small, 2013, 9 (1): 115-123.

[443] Begum P, Fugetsu B. Phytotoxicity of multi-walled carbon nanotubes on

red spinach (*Amaranthus tricolor* L) and the role of ascorbic acid as an antioxidant [J]. Journal of Hazardous Materials, 2012 (243): 212-222.

[444] Chen L Y, Yang S N, Liu Y, et al. Toxicity of graphene oxide to naked oats (*Avena sativa* L.) in hydroponic and soil cultures [J]. RSC Advances, 2018, 8 (28): 15336-15343.

[445] Hamdi H, De La Torre-Roche R, Hawthorne J, et al. Impact of non-functionalized and amino-functionalized multiwall carbon nanotubes on pesticide uptake by lettuce (*Lactuca sativa* L.) [J]. Nanotoxicology, 2015, 9 (2): 172-180.

[446] Wu X, Zhu L Z. Evaluating bioavailability of organic pollutants in soils by sequential ultrasonic extraction procedure [J]. Chemosphere, 2016 (156): 21-29.

[447] 宿程远,李伟光,刘兴哲,等.响应曲面法优化制备改性海泡石负载纳米铁材料的试验研究[J].环境科学学报,33 (4): 985-990.

[448] Varanasi P, Fullana A, Sidhu S. Remediation of PCB contaminated soils using iron nano-particles [J]. Chemosphere, 2007, 66 (6): 1031-1038.

[449] Zhang W X, Wang C B, Lien H L. Treatment of chlorinated organic contaminants with nanoscale bimetallic particles [J]. Catalysis Today, 1998, 40 (4): 387-395.

[450] 何绪生,耿增超,佘雕,等.生物炭生产与农用的意义及国内外动态[J].农业工程学报,2011,27 (2): 1-7.

[451] 郭延辉,樊静,王建玲,等.树脂固载纳米铁对偶氮染料直接湖蓝5B的脱色性能研究[J].环境工程学报,2010,4 (2): 337-341.

[452] 朱慧杰,贾永锋,姚淑华,等.负载型纳米铁吸附剂去除饮用水中As (V) 的研究[J].环境科学,2009,30 (12): 3562-3567.

[453] 马剑华.纳米材料的制备方法[J].温州大学学报,2002,2 (15): 79-82.

[454] Li X Q, Brown D G, Zhang W-X. Stabilization of biosolids with nanoscale zero-valent iron (nZVI) [J]. Journal of Nanoparticle Research, 2007, 9 (2): 233-243.

[455] Du J J, Zhou Q X. Preliminary study on effects of nano-scale amendments on hyperaccumulator indian marigold grown on co-contaminated

soils [C]. Proceedings of the 3rd International Conference on Energy and Environmental Protection，ICEEP，2014（955-959）：243-247.

[456] Hummers Jr W S，Offeman R E. Preparation of graphitic oxide [J]. Journal of the American Chemical Society，1958，80（6）：1339.

[457] 李权龙，袁东星. 替代物和内标物在环境样品分析中的作用及应用[J]. 海洋环境科学，2002，21（4）：46-49.

[458] 高园园，周启星. 纳米零价铁在污染土壤修复中的应用与展望[J]. 农业环境科学学报，2013，32（3）：418-425.

图书在版编目（CIP）数据

氧化石墨烯与植物修复系统交互作用的研究 / 杜俊杰著. —北京：中国农业出版社，2022.9
ISBN 978-7-109-29983-2

Ⅰ.①氧… Ⅱ.①杜… Ⅲ.①石墨烯－土壤污染－交互作用－植物－生态恢复－研究 Ⅳ.①X53

中国版本图书馆 CIP 数据核字（2022）第 167021 号

氧化石墨烯与植物修复系统交互作用的研究

中国农业出版社出版
地址：北京市朝阳区麦子店街 18 号楼
邮编：100125
责任编辑：史佳丽　魏兆猛
版式设计：杨　婧　　责任校对：吴丽婷
印刷：中农印务有限公司
版次：2022 年 9 月第 1 版
印次：2022 年 9 月北京第 1 次印刷
发行：新华书店北京发行所
开本：880mm×1230mm　1/32
印张：7.25
字数：202 千字
定价：55.00 元

版权所有·侵权必究
凡购买本社图书，如有印装质量问题，我社负责调换。
服务电话：010-59195115　010-59194918